C000231084

BLOODY MARSH

BLOODY MARSH

A seventeenth-century village in crisis

Peter Warner

with photographs by Nick Catling

WINDgather
PRESS

For John Barwick and friends ...

Bloody Marsh

Text copyright © Peter Warner, 2000
Photographs copyright © Nick Catling, 2000

Peter Warner has asserted his right under the Copyright, Designs
and Patents Act 1988 to be identified as the author of this work.

All rights reserved. No part of this publication may be
reproduced, stored in a retrieval system, or transmitted in any
form or by any means (whether electronic, mechanical,
photocopying or recording) or otherwise without the written
permission of both the publisher and the copyright holder.

Published by: Windgather Press, 31 Shrigley Road, Bollington,
Macclesfield, Cheshire SK10 5RD, UK

Distributed by: Central Books, 99 Wallis Road, London E9 5LN

British Library Cataloguing-in-Publication Data
A catalogue record for this book is available from the British
Library

Library of Congress Cataloguing-in-Publication applied for

Bloody Marsh: A seventeenth-century village in crisis,
by Peter Warner

ISBN 0 9538630 1 8 paperback

First published 2000

Typeset and originated by Carnegie Publishing Ltd,
Chatsworth Road, Lancaster
Printed and bound by Bookcraft (Bath) Ltd

Contents

List of maps and figures

Glossary

aqua vitae	a distilled alcoholic drink
cade	a measure of 600 herrings contained in a willow basket or cadebow
chapelry	a dependent church with limited sacramental rights usually served by a chaplain
copyhold	a type of tenancy registered on manorial court rolls a copy of which is held by the tenant
cullet sheep	sheep belonging to tenants but included in the lord of the manor's flock as part of a foldcourse
Declaration of Indulgence	royal declarations (1662, 1672, 1687–88) repealing repressive legislation against religious nonconformists and Roman Catholics.
demesne	land owned and cultivated directly by a manor or lordship
distraint	the seizure of goods in lieu of debt
doles	portions of land temporarily allotted for grazing, hay making or cultivation – usually allotted to poorer tenants
entry fines	payment made by a tenant to the lord of the manor on entering into a tenancy
fold-course	the legal right to graze and fold a flock of sheep over a specified area of land
glebe land	land in the ownership of the church farmed by a parish priest
intake	land temporarily converted from open rough pasture for enclosed agricultural use
messuage	legal term for a dwelling house with outbuildings, land and appurtenances assigned to its use
neatherd	man responsible for a herd of neats or heffers – castrated bull-calves
quit-claim	acknowledgement not to pursue legal rights or to proceed further in a dispute
rental	a document listing all the lands on a manor and the rents due from them
right of commonage	legal ownership of common grazing rights

Glossary	*royal patent*	a monopoly granted by royal charter in return for services or cash payment
	Sandling	a coastal region of East Suffolk with distinctive light sandy heathland soils
	sexton	parish officer responsible for maintenance of the church fabric, bells, churchyard and graves
	sheepwalk	area of enclosed rough pasture for grazing sheep
	ship money	tax imposed on coastal counties to finance warships – when this was extended to inland counties in 1635 it became a symbolic cause of grievance
	tithes	a tax of one tenth of the annual produce of land or labour such as fishing paid to the church
	wether	a castrated ram
	whin bushes	generic for furze or gorse (*Ulex europaeus*) or more specifically Petty Whin (*Genista anglica*)

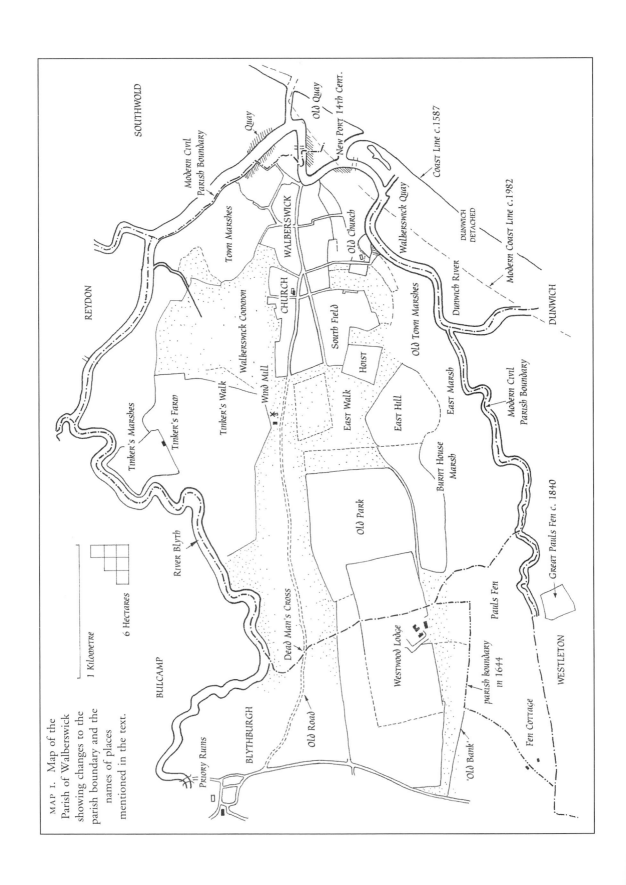

MAP I. Map of the Parish of Walberswick showing changes to the parish boundary and the names of places mentioned in the text.

1 Kilometre

6 Hectares

BULCAMP

REYDON

SOUTHWOLD

Priory Ruins

BLYTHBURGH

River Blyth

Tinker's Marshes

Tinker's Farm

Tinker's Walk

Modern Civil Parish Boundary

Quay

Town Marshes

WALBERSWICK

Walberswick Common

Old Church

CHURCH

Old Quay

New Port 14th Cent.

Walberswick Quay

Coast Line c.1587

DUNWICH DETACHED

Modern Coast Line 1982

Dunwich River

South Field

Hoist

Old Town Marshes

DUNWICH

Wind Mill

East Walk

East Hill

East Marsh

Modern Civil Parish Boundary

Burnt House Marsh

Dead Man's Cross

Old Park

Old Road

Westwood Lodge

parish boundary in 1644

Pauls Fen

'Old Bank'

Great Pauls Fen c. 1840

Fen Cottage

WESTLETON

Preface

In Thomas Gardner's account of Dunwich, written in the mid-eighteenth century, there is mention of an incident that happened a hundred years earlier, which involved Sir Robert Brooke, lord of the manor, who:

> set up a boarded house for men and dogs, near Pauls-Fenn, to keep out and drive away any cattle belonging to the town of Walberswick; when one of the keepers came into the said town, and quarrelling with the townsmen, a lamentable fray ensued in which four men lost their lives, which gave occasion for calling the fenn afterwards, BLOODY MARSH. (Document 13).

Gardner was using the Walberswick church wardens' account books, which he annotated. It is clear from other entries in these account books that this incident took place in April or May of 1644, during the height of the Civil War. It seems that following the death of one of Sir Robert's henchmen from wounds received in a fight, three Walberswick men were hanged for murder, even though the murdered man had clearly caused the affray. Unfortunately the county assize records for this period do not survive, but both the church wardens' accounts and a rich archive for the Brooke family among the Cockfield Hall papers do; it is these two sources which enable the story of *Bloody Marsh* to be told.

I first came across the Cockfield Hall papers in 1976 as a post-graduate student working in the Suffolk Record Office (SRO) on a study of medieval settlement in East Suffolk, completed as a Ph.D. in 1982. Some of the documents were still held by the Blois family, who very kindly allowed the SRO to have microfilm copies made so I could study them. Sadly, the family has now left Cockfield Hall, but the documents are retained at the Ipswich Branch of the SRO. We all owe the Blois family a debt of gratitude that this archive has been saved for the county.

Only some of the documents were relevant to my work at that time, but it was obvious that here lay an important quarry of historical material of great depth and complexity. Over the next twenty years I gradually transcribed parts of the archive while working on other projects. The focus has always been on the primary local sources; the existence of relevant national archives have been noted but not explored. Other documents may well come to light in government sources such as the Exchequer Depositions and the Court of Requests. There will always be scope for further research in national archives, but the purpose of this exercise was to chase the plot through the local sources searching for a balance between documents generated by 'government' and the 'governed'. Too often it is those in government who create the historical archive – here was a chance to redress the balance. My history students at Homerton became familiar with them in history workshops and mock courts attempting to prosecute and defend the actions of William Turrould and Sir Robert Brooke: a major stimulus for this book. I hope that the published sources in the appendix will continue to be used in this way for student history workshops.

I am indebted to the Suffolk Record Office for their patience and assistance over many years and the Trustees of Homerton College who kindly allowed me a term of study-leave to complete the work. Homerton also provided a small research grant for Barbi Danes who typed the appendix from my transcripts. Nesta Evans has been particularly helpful for her astute historical comments. Mary Martin has assisted by combing through the text. Dr David Smith and Professor John Morrill of Selwyn College have offered valuable comments. Richard Purslow has maintained his faith in my work and has encouraged me to do things that other academic publishers would never consider.

There are foundation stones which underpin this book: Rachel Lawrence's work on the eighteenth-century history of the Blyth, *Southwold River*, is essential reading for anyone interested in the later history of this area, while my own Ph.D. and provides the early historical background. Two other foundation stones deserve mention: the work of the Reverend Lewis, who transcribed the first churchwardens' account book and published it in 1947, and the recent work of M. E. Allen who transcribed the seventeenth-century wills for the Archdeaconry of Suffolk.

Bloody Marsh has been a significant departure for me from my usual focus on earlier periods of history. It has been a useful historiographical exercise, combining as it does the physical landscape with social and economic history. It is a strongly held view of mine that society and landscape are inseparable and that you cannot interpret the one without the other and *vice versa*. Archaeologists and anthropologists have long recognised that human beings and landscape interact with one another in such a way that the landscape can be used as a source for recognising major social and economic change, particularly so called 'revolutions' in the prehistoric and modern landscape. The enclosure movement in the early modern period was just such a landscape 'revolution'; it is important therefore that we use the seventeenth-century landscape as a source for understanding social change just as we would for earlier periods.

Nick Catling's photographs in many respects make the book. His work allows the reader to participate in the modern landscape and to appreciate fully its historic significance. Primarily this is a work of social and landscape history, but we expect some people will buy the book first and foremost for its photographs, regarding the text as an explanation for them. *Bloody Marsh* is therefore intended for a wide audience, historians and non-historians alike. Its structure of alternating long and short chapters, moving from analysis to narrative and back again, means that the story can be read by skipping the longer descriptive chapters if the reader so desires. Secretly I long for it to inspire others working in more imaginative art forms, a historical novel perhaps, a film, better still, like Peter Grimes, it would make a wonderful opera. History has always been used creatively, for good or ill; it is the duty of historians to encourage its usage for good particularly among the young.

Of course the story is open to re-interpretation – that is why the primary sources have been published with it. Here we have just one interpretation; I reserve the right to change my interpretation in the light of any new evidence, with time and deliberation. *Bloody Marsh* is intended primarily for enjoyment, it is a ripping good yarn, and as such I hope it will also be recognised as good history. All the errors of fact and imperfections of style are mine alone.

Peter Warner   Cambridge 2000

FIGURE 1.
Marsh thistle
(*Carduus palustris*).

CHAPTER ONE

The prelude

There the blue bugloss paints the sterile soil;
Hardy and high, above the slender sheaf,
The slimy mallow waves her silky leaf;
O'er the young shoot the charlock throws a shade,
And clasping tares cling round the sickly blade;
with mingled tints the rocky coasts abound,
And a sad splendour vainly shines around.

Crabbe

In 1644, two years after the outbreak of the Civil War, there was an affray in the town of Walberswick caused by one of the landlord's men, 'a stout fellow', who started a fight with the townsmen over grazing rights. The argument stemmed from a long-standing dispute over marshland enclosure. As a result of this incident one man died and three of the townsfolk were subsequently hanged for his murder. The circumstances which led to this tragedy were recorded in detail by one of the churchwardens, John Barwick (Document 14); he also noted that the disputed area was thereafter known as 'Bloody Marsh'. Barwick kept records in the churchwardens' 'ancient books' which were to prove vital legal evidence in the protracted dispute over enclosure. These books were later used by the antiquarian, Thomas Gardner, in his historical account of Dunwich published in 1754. Legal documents relating to the enclosure of the commons in the sixteenth and seventeenth century also survive among the Cockfield Hall papers, once the archive of the Brooke family, lords of the manor of Blythburgh *cum* Walberswick. Thus evidence for two sides of the argument survive for posterity. Both the Cockfield Hall papers and the churchwardens' account books are now preserved in the Suffolk Record Office at Ipswich; these two collections provide the primary historical source for this book.

On the face of it this was a trifling incident in national terms, one of countless known and unknown fracas over enclosure that took place all over England in the seventeenth century. Similar struggles were taking place in the West Country and in the Fenlands of East Anglia. Closer to home, in 1642, there were anti-Catholic riots in and around Colchester and the Stour valley. In 1629, there were dearth riots over the export of grain at Maldon, where executions followed repeated acts of violence. The *Bloody Marsh* story itself,

FIGURE 2.
Panorama from the
top of Westwood
Lodge tower. Here Sir
Robert Brooke would
have commanded a
view out over his
formal gardens in the
foreground below with
Pauls Fen, in the
distance on the right.
The area of 'Bloody
Marsh' in the centre
distance. East Marsh
can be seen on the
extreme left. On the
horizon, Dunwich
Forest, which would
have been open
heathland in the
seventeenth century.

like similar local flash points sparked off by long-standing social problems such as scarcity of food, religion and enclosure, appears isolated and essentially unrelated to the events of the Civil War. However, civil disorder and the inability of men to find satisfaction through the processes of law were arguably contributary factors behind the Great Rebellion; Bloody Marsh was not an act of rebellion in itself although there were rebellious clashes, particularly in the relationship between customary tenants and the squire. The story illustrates the great complexity of local issues and loyalties, and highlights the build-up of economic stress in a local community prior to the breakdown of law and order – something which can happen anywhere at any time – but which also contributed to the Civil War.

The causes of the Civil War are a source of on-going debate. In the 1960s and 1970s the idea of 'localism' was prevalent, in particular the work of Alan Everitt on the counties of Suffolk and Kent suggested that local issues were sometimes more important, particularly at a county level, than the national issues which concerned centralised government. More recently the work of Clive Holmes and others have highlighted the differences between regions and the complex social and political make-up, varying substantially some-times within counties. Blackwood has noted significant differences in the seventeenth-century political makeup between East and West Suffolk. The enclosures which led up to the *Bloody Marsh* incident were clearly a local issue which served to divide the community, but much of the legal action took

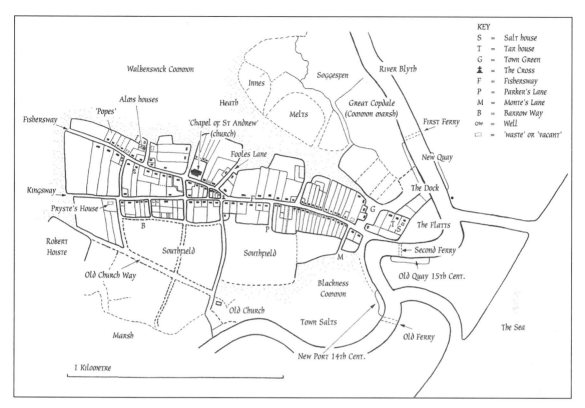

KEY

S	=	Salt house
T	=	Tar house
G	=	Town Green
♅	=	The Cross
F	=	Fishersway
P	=	Parker's Lane
M	=	Monte's Lane
B	=	Barrow Way
ow	=	Well
⌑	=	'waste' or 'vacant'

Walberswick Common · Soggespen · River Blyth
Innes
Heath · Melts · Great Copdale (Common marsh) · First Ferry
Alms houses
'Popes'
Fishersway · 'Chapel of St Andrew' (church) · New Quay
Fooles Lane · The Dock
Kingsway
Pryste's House · G · The Flatts
B · P · S
Robert Hoiste · Second Ferry
Southfield · Southfield · M · Old Quay 15th Cent.
Old Church Way
Blackness Common
Old Church · The Sea
Marsh · Town Salts · Old Ferry
New Port 14th Cent.

1 Kilometre

MAP 2.
A visualisation of the town of Walberswick based on a written survey of 1582/3 and the modern Ordnance Survey map. Details from Ralph Agas' map of Dunwich in 1587 have been incorporated (see figure 10).

place in Whitehall, London, where the squire was well connected. Walberswick itself was also connected to a wider world via the coastal trade and was more cosmopolitan than the average rural community. So although *Bloody Marsh* may seem to be a localised dispute it can only be fully understood in relation to its wider social and political context of England in the mid-seventeenth century – the period of the Civil War.

More remarkable is the depth of detail in the primary sources, carefully selected and abbreviated here, and the colourful personalities revealed by them. The modern trend in local studies is towards 'total reconstitution' of a community from documentary sources. Some readers will be disappointed because that has not been attempted here; unfortunately the parish registers do not begin until 1656, well after the events in question. The rentals and surveys on the other hand allow for a detailed reconstruction at certain dates. The best survey is a manorial rental of 1582/83 reconstituted here as Map 2; nine others survive, some fragmentary, for various years between 1601, and 1757. In order to reconstruct the town from the rental of 1582/83, the closely written text of which extends over 39 pages, it was necessary to piece together a gigantic verbal jig-saw puzzle. The result gives an approximation based on certain fixed points, such as the church, the Green and surviving roads and lanes (Map 2). It is not an attempt to reconstruct the exact land area of each plot and must not be taken literally in this respect. It does, however, give us a fixed point from which we can visualise the town towards the end of its hey-day in the late sixteenth century. A few deserted tenements are evident, but the impression given is of a populous, thriving community focused on its church and quayside. In theory it would be possible to reconstruct the town at other points in time where there are surviving rentals, but in practice the rentals are very repetitive, being dependent on descriptions contained in the court rolls, and little would be gained by such an exercise. Likewise, the complete transcription of all the documents including the sixteenth and

FIGURE 3.
Walberswick green and the village sign familiar to holiday-makers and artists who have frequented the village since the late nineteenth century. The 'Town Green' is described in a survey of 1582–83.

FIGURE 4.
The north corner of
Walberswick green.
The house on the left
stands on the site of
tenements held by
Thomas Pryme and
Thomas Freeman in
1582–83.

seventeenth-century church-wardens' books would be a massive project, interesting and desirable in itself, but they would probably add very little to this story. To get this far it has been necessary to single-mindedly chase the story through the archive deliberately leaving out material which other historians might regard as important. Perhaps this book will be the stimulus for further research.

One might suppose that both parties in the dispute over enclosure in the seventeenth century would fall neatly into the two opposing camps of the Civil War, but that is not the case. Readers must suspend all their preconceptions about the period, for that is the whole point of this book. Bloody Marsh is not a simple story of Cavaliers and Roundheads; the religious beliefs of both parties were not dissimilar as far as such opinions can be ascertained from the surviving material. It was an argument amongst puritans in what was a largely puritan social climate. However, the 'modernity' of puritan thinking, on a more rational level, and an underlying concern for basic freedoms which were undoubtedly imbedded in it, underpin the legal battles over enclosure. Both parties were arguing for the *ancien regime*: attempting to re-establish what they regarded as their basic freedoms. On the one hand the freedom to exercise free will and enclose one's own land, and on the other the freedom to practise ancient customary grazing rights. Religious thinking therefore helped both sides to focus their arguments and substantiate their claims, just as it did nationally in the Civil War.

Although the area was relatively untouched by the military exchanges of the Civil War, it was touched by local events that related to the conflict. We hear of troops being mustered in the neighbouring parish of Blythburgh, of the visit of William Dowsing the iconoclast and his troopers to strip the church of superstitious images and inscriptions. We hear also of a large sum of money being collected, inspite of almost overwhelming poverty, to assist Cromwell's campaign against the 'Irish rebels'. The story reveals the great complexity of social and economic issues at a local level. Nothing can be taken for granted; society was probably just as complicated then as it is now, and people do not fit into water-tight compartments. Local issues often have the potential to turn national trends on their head; or they may under-score political issues adding bitterness to debate and unwarranted loyalty between individuals where it is least expected.

Local history only makes sense when it is related to a wider world of social, political and economic change, and it may not always fulfil our expectations. Just occasionally there is sufficient detailed documentation to know what some people thought was happening to their communities at a particular time, though we cannot expect their opinions to be without prejudice. In the story behind *Bloody Marsh* there is enough material to give a multi-faceted account of what was happening to a particular village on the east coast of England – Walberswick in Suffolk – its society and its landscape – in the seventeeth century and how people reacted to a changing world. We do not get the whole picture, but we get just enough pieces to establish a meaningful interpretation. The fulcrum of these events was a local struggle over enclosure, of little importance then nationally and of little importance now. But its relationship to national, political, economic and governmental changes at the time is fascinating, so too are its colourful protagonists. It is a story full of drama, of human weakness as well as human strength; as such it is a story worth telling.

In essence *Bloody Marsh* is a history of enclosure and the poverty and distress found in some sixteenth-century rural communities: as such it is nothing new. The wider economic background of demographic change, rising food prices and economic inflation was an important stimulus to enclosure; it was also a stimulus to increase the value of rents and a cause of poverty as wages failed to keep pace with inflation. The consolidation of landholdings that followed enclosure led to a polarisation within the farming community as larger farms prospered and smaller ones failed or sold up. Socially and economically it was a painful process; for some it brought prosperity, for others it brought ruin.

Enclosure has a long history in East Anglia and it was often associated with violence and rebellion; Kett's Rebellion of 1549 was still within living memory in the early seventeenth century. Although Kett himself was an encloser, his rebellion was essentially about enclosure of the common fields and declining wages. The story of *Bloody Marsh* is about the enclosure of saltmarshes and barren heaths; the sort of landscape which can still be seen along the Suffolk coast today. This landscape forms the essential backdrop to the dramatic events leading up to conflict; it is the social interaction with this landscape at a time of fundamental change that is so interesting.

There is no doubt that enclosure was a contributary factor in creating poverty at Walberswick. But enclosure here was inflicted on a community that was already experiencing severe economic decline, as was often the case elsewhere. Enclosure was often the one issue where villagers were prepared to stand up and fight for their rights, as they did with such vigour at Walberswick. However, enclosure was not the prime cause of poverty. It is important to remember that there were also many farmers and labourers who benefited from the intensification of agriculture that followed on from enclosure. The resulting national increase in food production, particularly from the enclosure of fenland, resulted in a reduction in rising food prices in the second half of the seventeenth century, at a time when the population increase was also levelling off and there was greater economic stability. Enclosure enabled a rising tide of population to be fed, particularly in the larger towns and cities, and it is now seen as one of the prerequisites of the Industrial Revolution. It is also important to remember that the process whereby enclosure was achieved sometimes resulted in loss of income to the landowner; great expense was wasted in legal costs, fines and damages, not to mention the cost of draining, embanking, fencing and constructing sluices. Success was not always guaranteed, indeed some landlords ruined themselves in the process.

Walberswick was originally a coastal fishing community, but following the decline of the fishing industry in the years after the Dissolution of the Monasteries, it turned to the coastal trade in butter and cheese. By the seventeenth century its small harbour was too shallow to take the larger vessels then being built, so its efforts to satisfy the demands of coastal traffic met with limited success. As these sources of income dried up, so the townsfolk turned to agriculture for subsistence, but their landholdings were pathetically small and the soil desperately poor. So they came under pressure to graze their heathland commons and marshes more intensively. Here they met the lord of the manor head on. The Brooke family had purchased the land in the late sixteenth century and immediately set about banking and draining the fens and marshes, and converting tracts of heathland into enclosed sheepwalks. The townsfolk could do little about the collapsing fishing industry, or the destructive affects of the sea on their harbour. Likewise, rising food prices and the inflationary effects on wages were as much out of their control as the dreadful heathland fires that periodically raged through their houses. But they *could* fight a greedy and belligerent landlord, and this was where their efforts and frustrations were acted out, through litigation, through local politics, and eventually through physical violence.

The history of enclosure at Walberswick is as complex as it is anywhere else, but to understand the circumstances that led to the physical confrontation over *Bloody Marsh* we need to go back to the late middle ages. In 1430, John Hopton, 'a youth of great expectations', inherited the vast Swillington estates in Yorkshire and Suffolk. Ten years later he consolidated his estates in Suffolk by purchasing the manor of Cockfield Hall in Yoxford, the only manor in Yoxford not already part of the Swillington estates. Although he was a Yorkshire man and his estates there were probably worth more than all his Suffolk estates put together, he chose to live at Blythburgh in his manor house of Westwood. The subsequent management of the Hopton estates forms the background to the enclosures of the sixteenth and seventeenth centuries.

Colin Richmond has published a brilliant analysis of the life of John Hopton and his estates, so they need only be touched upon here. Suffice it to say that Hopton was a resident squire (without political ambitions so far as we know), who took some personal interest in his newly acquired estates. There is evidence that the demesnes attached to the manor of Westwood were gradually extended in the fifteenth century, mostly as sheep pasture run as two fold-courses. After John Hopton's death in 1478 his estates passed in rapid succession through his two sons, Thomas and William, to his grandson Sir Arthur Hopton who inherited in 1499. The next two generations of Hoptons, unlike their predecessors, followed political careers and in the early sixteenth century there was a need to lease out the demesnes and the fold-courses for cash.

In 1534, Reginald Barnard, described as 'gentleman', Robert Pygott 'merchant', William Barrett 'yeoman', Robert Burford 'mariner', and John Tompson 'cordwainer' (leather merchant), mostly Walberswick men, entered into a ten-year lease with Sir Arthur Hopton to rent his sheep-pasture for £20 per annum. The various occupations of these men suggest a range of skills and business acumen essential to the wool and leather trades. They took the sheep-pasture between Blythburgh and Walberswick together with a close called Feld Close lying in Blythburgh and another called New Close. They also had the use of the barn within the park at Westwood: 'therin to lay ther corne together with the chawmbyrs ther for to leye in ther wolle and ther hors harnes' (Document 1); there had been a woolhouse at Westwood in 1478. They could take alders from the aldercarr called Borwards or Mekylfyldes in Blythburgh to make bridges for their 'shepyes gate' into the fen. Payment was to be made half yearly at Easter and Michaelmas. Sir Arthur or his heirs could give twelve months' notice if he wished to close on the lease or if he wished to, 'leye it with his own catell or theres.' So the townsmen of Walberswick became the farmers of the lord's sheep-pasture extending across the parish boundary into Blythburgh.

This situation continued for just seven or eight years, then in 1540/41 Sir Arthur Hopton and Owen Hopton his son required the townsmen to give up their lease, to their 'great losse and hyndtrans', in favour of a neighbouring

FIGURE 5.
The southern
boundary bank of
Westwood Park near
Hoist Wood. The
park was in existence
in the fifteenth
century and may be
much earlier. The
Brookes split up the
park in the sixteenth
century in order to
lease out the land to
tenant farmers.

landowner, Sir Edmund Rous, who then: 'did peaceably and quyetly enjoye posses manure and feide the sayed walk and sheipes pasture; by the space of fyve yeres and more.' Rous ploughed some of the land and sowed it with barley, but then a dispute arose between him and the townsmen. Cattle belonging to the men of Walberswick were impounded by Rous's servants as they grazed towards his barley: 'for the damages wherof, such unreasonable charges were answered contentyd and payed to the sayed Syr Edmund Rous hys servants as they would require …' (Document 2).

The dispute was such that it caused a rift between the two landlords. In fear of forfeiting the land because the terms of the lease had been broken, Sir Arthur and Owen Hopton bought back all the sheep together with the lease. In 1556 there was an attempt on the part of the town to recover the lease for themselves, or at least some of the grazing rights. Seventeen of them signed a lengthy statement which declared that they knew what had happened, perhaps in an attempt to curry favour with the Hoptons and absolve themselves from blame (Document 2).

These two documents illustrate a number of problems emerging in Blythburgh and Walberswick in the late sixteenth century. Firstly there was increased competition for grazing resources. Secondly, some enclosures, particularly the sheep walks, had already taken place and others, intakes from the extensive heathlands, were just happening. Thirdly, some pasture was being broken up by the plough in addition to that which was normally ploughed as part of the fold course. Lastly the lease itself illustrates the growing interdependence between upland sheep-pasture and lowland, marsh-land grazing. There was also a notable lack of definition of the areas covered by the leases to which common rights applied. So the struggle for control of the sheep-pastures and commons had begun before the manor was purchased by the Brooke family in 1592, and long before the more serious confrontations of the seventeenth century.

The prelude: chapter notes

Extract from George Crabbe's poem, 'The Village' (lines 72–8). For this and subsequent extracts from Crabbe see: Poster (ed.) 1986. The Walberswick churchwardens' accounts (CWA) for the years 1582–1700 can be found in the Suffolk Record Office (SRO) at Ipswich: FC 185/E1/2; the years from 1450 to 1499 were edited and published by R. W. M. Lewis, 1947. See also Gardner 1754 (Gardner). For comparable incidents see: Sharp 1980; Underdown 1985; Walter & Wrightson 1976, and the more recent work Morrill & Walter 1997. Fenland riots: Lindley 1982. Anti-Catholic riots: Walter 1996. For 'localism' see: Everitt 1960 and 1966; Holmes 1997; Blackwood 1997. Local history has made a significant contribution to our understanding of the origins of the Civil War, see: Richardson (ed.) 1997. For the importance of religion as an underlying cause of the Civil War see: Morrill 1984. The 'extent' or survey of Walberswick in 1582–83, which makes no reference to rent, survives in the Ipswich record office: SRO: HA30: 50/22/10. For all subsequent rentals and surveys: 1610 (HA30: 50/22/3.22); 1612–27 (/8.1); 1635 Blythburgh & Walberswick (/10.9); 1645 (/8.3); 1694 (/8.17). There are also rentals for 1694, 1736, 1738 and 1748–57. To avoid confusion the term 'rental' has been applied throughout in the text. For the Hoptons see: Richmond 1981; reference to the 'woolhouse' *ibid.*, p. 36. William Dowsing visited Walberswick on 8 April 1644: Gardner, p. 160; White 1883/88. Sources for social and economic change in the seventeenth century are discussed in chapters 2 and 4. For the biography of John Hopton, see: Richmond 1981.

CHAPTER TWO

Poverty and decline

....

I grant indeed that fields and flocks have charms
For him that grazes or for him that farms;
But when amid such pleasing scenes I trace
The poor labourious natives of the place,
And see the mid-day sun, with fervid ray,
On their bare heads and dewy temples play;
While some, with feebler heads and fainter hearts,
Deplore their fortune, yet sustain their parts:
Then shall I dare these real ills to hide
In tinsel trappings of poetic pride?

Crabbe

In 1450, the townsfolk of Walberswick paid for the topping out of their magnificent new church tower; the work had been commissioned twenty-five years earlier. Eight wind-vanes costing 4s. 9d. glittered on the pinnacles of the steeple; another 10s. was paid to cover them in gold leaf. This day marked a climax in the town's history, for Walberswick was then in every sense a town, a minor sea port with a thriving quayside. In 1467 a set of bells was added to the tower costing £26 – a small fortune by the standards of the day. Of course such extravagance was a necessary part of competitive church building in the fifteenth century. Other communities with similar towers at Covehithe, Blythburgh, and Dunwich, were put to similar expense. But one hundred years later it was a very different story, for in 1585 the churchwardens of Walberswick were forced to sell their great bell to pay for repairs to the church. The town was then described as 'very poor'. Poverty worked like a slow debilitating disease sapping both the community of its spirit and robbing its buildings of their finery. Yet the causes of poverty at Walberswick were complex and need to be carefully considered.

In the late Middle Ages Walberswick's position on the coast and its relationship with the river Blyth was geographically unstable; its economy constantly lurched from sea to land and back again. This inherent instability caused untold distress and hardship. The movement of church and settlement amply illustrates this fact as the town and quay followed the changing course of the river and its outlet to the sea. As the harbour mouth and river moved so quayside and settlement followed hard behind, for the harbour was the

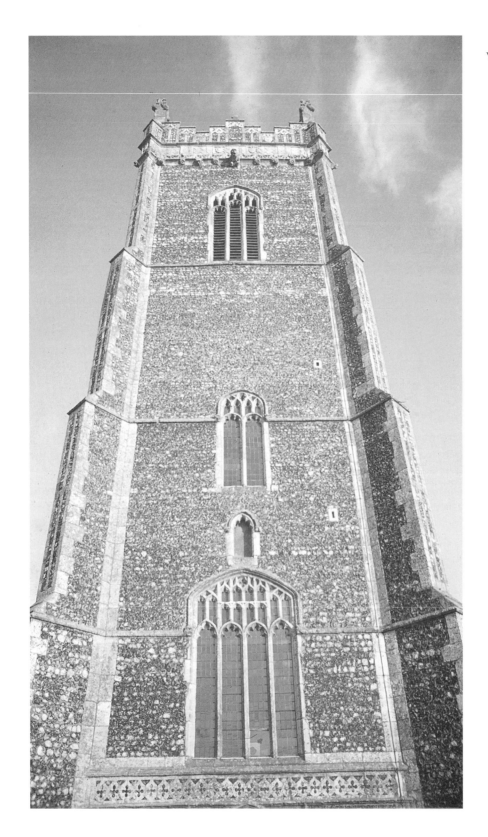

FIGURE 6.
Walberswick church
tower.
PETER WARNER

Poverty and decline

FIGURE 7.
Walberswick and the
river Blyth looking
north towards Black
Shore with the town
of Southwold in the
distance. Fishing has
always been the staple
industry here,
although it has now
been overhauled by
tourism. From here in
the sixteenth century
butter and cheese
were exported to
London.

PETER WARNER

heart of the community. As new quays came into use so the older ones fell into decline. The quay was the platform over which passed not just the merchandise of Walberswick, but of all the surrounding district.

The town's economy had four basic constituents, farming, fishing, ship-building and coastal trade. Each of these had the potential to be highly lucrative, they also had the potential for considerable economic diversity, but all of them were subject to national economic trends. There is evidence for some flexibility on the part of the townsfolk, many of whom had fingers in different pies, particularly farming and fishing, but these were seasonal and subject to market fluctuations. Coastal-trading, transporting goods to larger seaports, could always be done out of season, or on the side. But the small-scale nature of its business, which was always done in close competition with neighbouring coastal ports, and the geographic circumstances of Walberswick combined with national economic trends to create brief flourishes of prosperity, followed by long deep troughs of economic decline.

Agriculture at Walberswick depended not only on market cycles, but also on the meagre returns from its poor sandy upland soils. Its low lying marshland pastures could provide good rough grazing, but they were subject to inundations from high salt water tides. Fishing and trade were heavily dependent on the state of the harbour, which was constantly shifting, silting up, or unapproachable in bad weather. Ship-building was dependent to some extent on government contracts during times of war and at other times on

13

the unpredictable supply of merchant shipping. As the demand for vessels with larger draughts developed in the later Middle Ages, the limited width of the river and the shallow harbour entrance restricted the size of ships that could be commissioned there; deeper ports, such as Ipswich and Woodbridge were better placed to win ship-building contracts.

When wool prices were high and the monasteries were demanding a largely fish diet, as was the case in the late fifteenth century, Walberswick flourished along with other coastal towns. The monastery at Blythburgh took the highly lucrative tithes for herring and sprats landed at Walberswick – amounting to a fifth of the catch after deductions for labour, food and transport. These revenues assisted in the rebuilding of Walberswick church in the fifteenth century, a golden era in the town's history. This was also the time when the harbour mouth had shifted northwards from Dunwich into the parish of Walberswick and tolls could be demanded from vessels using the quay, although this was vigorously disputed by Dunwich.

In the mid-fifteenth century, when Walberswick barques were sailing for Iceland, Shetland and the northern seas, other parishes further inland were experiencing economic decline. Walberswick enjoyed some of its success at the expense of Dunwich during this time, but other, larger, coastal ports, in

FIGURE 8.
Fishing boats moored at Blackshore looking towards Walberswick in the distance. Fishing is still a significant local industry, just as it was in the sixteenth century.

14

FIGURE 9.
A Penzance fishing
boat moored on the
Southwold side of the
river Blyth. Long
distance trade and
fishing expeditions to
Iceland and the
northern seas were not
unusual here from the
fifteenth century.
Southwold had an
extensive port in the
sixteenth century,
which lay on the
opposite side of the
river to Walberswick
quay. Southwold's
church tower and
light house can be
clearly seen on the
horizon.
PETER WARNER

particular the boroughs of Ipswich, Orford, Aldeburgh and Yarmouth, were also prospering and some of their trade was rubbing off on intermediate coastal settlements, such as Kessingland, Southwold and Walberswick.

After the Dissolution of the Monasteries in the 1530s the fishing industry went into gradual decline. The wool trade also suffered because of wars on the Continent, particularly after the burning of Antwerp in the late sixteenth century, where much East Anglian wool was being exported. As the wool trade declined there was a shift towards coastal trading in butter and cheese to London, which was expanding rapidly in the late sixteenth and seventeenth centuries. By 1565 there were sporadic problems with pirates. In the 1620s there were losses to the 'Dunkirkers' and 'Hollanders'; so in 1627 it was necessary to provide a protective convoy for the fishing fleet heading north to Iceland.

The first clear indication of economic problems at Walberswick come in the 1580s when the lord of the manor, Sir Owen Hopton, ordered that the butter and cheese carriers should pay towards the maintenance of the church at the rate of 2*d.* a load. This order was supported by the signatures and marks of 29 townsfolk. Three reasons are given for this order. Firstly: 'The great decay of occupying of fishing-fare, which of late years have been frequented and used by this our poor town … by which trade the town was and is yet simply maintained.' The churchwardens declared that they could no longer maintain their various charges: 'without some other contribution

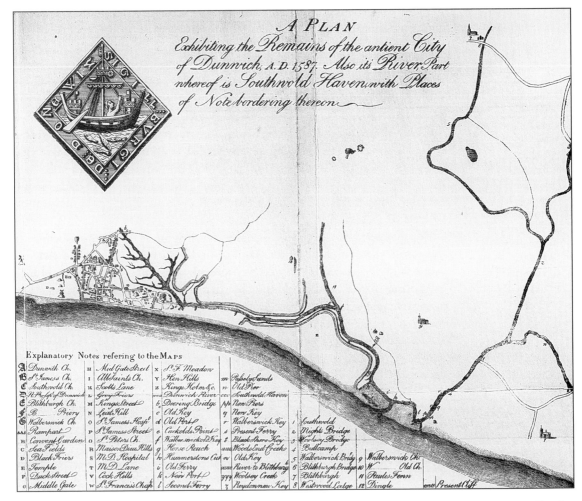

FIGURE 10. Detail from Ralph Agas's map of Walberswick and Dunwich dated 1587 as printed by Thomas Gardner in 1754.

or relief, had out of some other trade or traffic than fisher-fare.' Secondly: 'that divers good houses of late time, by misfortune of fire, are consumed.' Gardner lists five major fires which burnt parts of the town; the first took place sometime before 1583. Thirdly, it was stated that: 'our town is not now inhabited with so good occupiers, as it hath been in times past.' This may be no more than an indication that some wealthy occupiers had moved out, or had declined in wealth, a symptom rather than a cause of economic decline. Nevertheless it was seen at the time as contributing towards economic distress because the relief of the poor was dependent on the rates paid by wealthier citizens, without whom the poor could not be sustained.

Poverty was as endemic in the Tudor and Stuart countryside as it was in some towns. Gifts were frequently made in wills to assist the poor and it was the churchwardens and overseers of the poor who levied the poor rates and administered alms. In 1522 a small almshouse was given by the Odiron family it was split into three dwellings and appears in 'Almes Lane' on the rental of 1582/83 west of the church (Map 2). In 1572, Alexander Richardson, the head of a Walberswick peasant land-holding dynasty, gave his: 'houses in Walberswick sometime called Popes … to be an alms house … upon the condition that the townsmen there shall maintain the same in good repairation and pay the rent.' It too appears close by in the same rental, but is not mentioned as an almshouse, presumably because the rental was based on its old copyhold description. In the 1580s there are large numbers of payments in 'almes' listed in the church-wardens' accounts and a significant number of properties in the 1582/83 rental were held by widows while others were vacant. Perhaps the extensive commons were used for the relief of the poor and the associated poor rates, but the soil was so poor that it must have had limited effect.

It was recognised that the state of the harbour was critical to the economic survival of the community and efforts were directed towards its repair. In 1619 the churchwardens received £22 2s. 6d., which was delivered to them by Sir Robert Brooke on behalf of the Fishmongers' Hall in London who had responded to an appeal made by Dunwich, Southwold and Walberswick for 'repairing and making a haven.'

The sale of the great bell in 1585 was endorsed by an assembly of fourteen townsfolk; it weighed 1707 lbs and was bought by Thomas and George Smith of Ipswich for £26 8s. 9d. In the following year, the bishop's visitor required the churchwardens to repair several books and to provide others. They were

FIGURE 11. The footbridge across the old river channel which now gives access to the beach for holiday-makers, probably marks the position of a sixteenth-century ferry, which once gave access to the 'Old Quay' mentioned on the east side of the channel in 1582.

also asked to provide a new surplice and a cover for the communion cup, they were to mend the church-yard fences and gates; there is a hint of mismanagement and dilapidation. But the lack of a suitable surplice and cover for the communion cup might also indicate a puritan presence, or even a puritan clergyman; the strictures from the bishop might therefore indicate an attempt to find fault and tighten up on 'correct' forms of worship at Walberswick.

In 1597, at an assembly of 45 residents, the churchwardens were given authority to lease out the ancient sources of revenue; what amounted to a mortgage of future income. This included the herring and sprat fishing doles, the duties from ships returning from Iceland and the north seas (3s. 4d. for every voyage), the duty of 2d. a load on butter and cheese, the duties on, 'great beasts' grazing on the commons, the duty called 'kadge': a tax on every cade of 600 red-herring. No explanation was given for this drastic measure, but clearly there was a financial crisis and this was seen as a way round it.

In October 1609, tension between the declining numbers of fishermen and those increasingly occupied in the more lucrative coastal trade, particularly the butter and cheese carriers, came to a head. The levy of 2d. a load initiated in 1583 had done nothing to check the movement away from fishing. At Beccles quarter sessions court it was resolved that only the old men who had spent their former years in the fishing trade should occupy the coastal trade in butter. The young men were required to diligently attend upon the fishing craft, for: 'the neglect of the fishery was the means leading to the destruction of a nursery that bred up fit and able masters of ships and skilful pilots, for the service of the nation.' The fishing trade provided a vital supply of sailors to man the warships of a budding navy; some attempt was therefore made by Tudor and Stuart governments to encourage the catching and consumption of fish. So

FIGURE 12.
Walberswick Ferry. Tourists and walkers taking the ferry across the river Blyth to Black Shore, with the town of Southwold in the distance. Three ferries existed in sixteenth-century Walberswick and one of them was very close to this site.

FIGURE 13 (*opposite*). One of the old river channels looking towards Southwold. Here a ferry traversed a river bustling with commercial traffic in the sixteenth century. To the left stood the 'Tar house' of Alexander Richardson and the 'Salt house' of William Harborne in 1582–83.

19

just eight vessels were licensed to trade from Walberswick, taking it in turns without molesting one another and were bound over to keep the peace.

In October 1628, a warrant under the Poor Law was issued for the relief of: 'fourscore persons and upwards'. A petition had been raised sixteen months earlier, now thirteen neighbouring townships were ordered by the Justices of the Peace to contribute between 1d. and 6d. each a week to the churchwardens at Walberswick for relief of the poor. The rate was confirmed again in January 1629 and in April the same year a concerted effort was made to enforce payment of arrears with threats of distraint and jail for those who failed to pay. The 1620s was a time of dearth and other communities were having difficulty making ends meet; between the good harvest year of 1620 and the bad harvest of 1622 the national average price of wheat doubled from 23 shillings to 46 shillings a quarter.

It was said in the petition that the inhabitants of Walberswick: 'have long time made complaint to us of the misery and distressed estate of the poor people who are so many in number and so exceeding poor as the said inhabitants being most of them seafaring men and of very mean estates are no way able of themselves to levy suficient sums of money for their relief ... We do conceive that they or some of them are very like to perish for want of necessary food and sustenance, unless a speedy course be taken for a further supply therein.'

It is hardly surprising that under such circumstances vagrants were treated with great severity under the poor laws. On 23 May 1609, Jeffrye Purdie, an *aqua vitae* seller and fortune teller, was publicly whipped and sent back to his home town of Rye with his wife Mary. The distilling and sale of *aqua vitae* was commonly associated with Huguenot immigrants, escaping religious persecution on the Continent, who settled extensively in East Anglia, particularly in the major towns of Norwich and Colchester. Huguenot migration is also recorded at Ipswich and Dunwich from the late sixteenth century. These 'Strangers' were often subject to harsh and discriminatory laws. At Norwich where about six thousand of them had settled by the end of the sixteenth century, *aqua vitae* sellers were forbidden to sell their liquor on the streets.

Another serious fire was recorded in 1633; this time two supposed arsonists were arrested and escorted to Ipswich assizes. The churchwardens' accounts for 1633 report two men being sent by the appointment of the justices to bury the prisoners. It is possible that they had been executed, although death in prison from other causes, such as disease and violent behaviour is also likely. The rental of 1635 lists six 'burnt' properties and many others are described as 'waste'. The messuage of Anna Howlett fronting onto Fishersway in the middle of the town had been rebuilt after a fire, but Thomas Brocke had lost a building which had not been replaced and Isaac Burford had lost two tenements through fire, one on Fishersway. If one allows for other properties not included in the rental, we have a vision of a town that was only slowly recovering from a bad fire, the blackened remains of several buildings being still visible three years after the event.

FIGURE 14.
Wave gnarled timbers beside the old concrete harbour pier at Walberswick, playground of holiday-makers, they mark the position of the harbour entrance dug out by hand in the late sixteenth century.

FIGURE 15.
Groins now protect the coast at Southwold from coastal erosion.

Map 3, which is based on the rental of 1582/83, illustrates 104 dwellings of which thirteen were then vacant. Gardner, working from the churchwardens' accounts, says there were 71 'families' resident at Walberswick in 1633, but in the following year there were only 54. In that year he also says there were just 156 communicants. We cannot rely on these figures, but they allow an approximation of a declining population: 156 communicants divided by 54 households comes to a figure of 2.8 persons per household. If Gardner's figures are correct and seventeen households had disappeared from Walberswick between 1633 and 1634, then we may consider that at least 47–48 adults, nearly a third of the population, had departed.

Migration to New England is likely; Thomas Pell, a tailor from Walberswick left London for Massachusetts as early as 1635 as did John Read and his wife from Blythburgh. Fifteen persons left Southwold in the one year of 1637, including members of the Cocknan and Jeggles families. William Cocknan,

FIGURE 16. The River Blyth, the ferry stathe and Walberswick in the distance.

FIGURE 17. Extensive areas of saltmarsh once extended east of Walberswick, now only a few fragments remain as here looking towards the harbour mouth, with the caravan park and Southwold in the extreme distance.

mariner, made a second crossing that year with his wife, two children and two servants. Others, such as the Paine, Fiske and Thurston families left as part of the 'Wrentham movement' in 1637–38 following the inspiration of William Brewster who had sailed on the *Mayflower*. These migrations were largely organised by John Winthrop of Groton in Suffolk, who became first governor of the Massachusetts Bay Colony after June 1630. However, other less well documented settlements took place in the Caribbean on St Christopher's (St Kitts), which were developed from 1623 by Sir Thomas Warner of Parham and on Barbados after 1625. There was a movement to Antigua from Suffolk when Lord Francis Willoughby of Parham gained the lease of a royal patent to the Caribbean in 1647, he, together with Warner, organised migrants from Woodbridge and surrounding villages. Place-names in Antigua, such as Parham, Eye and Willoughby suggest a significant Suffolk migration. Later in the century there is evidence for migration to South Carolina from Westhall and Brampton, just five miles inland from Walberswick, involving the Gosnold and Bohun families. Edmund Bohun was settled as a merchant in Charles

Town before his father joined him, having been appointed Chief Justice of South Carolina in 1698.

When a report was made on the state of the town in 1654, the townsfolk were said to be impoverished, because of loss of commons, fire and death or removal of their ablest townsmen. The Hearth Tax Returns of 1674 are not wholly reliable, but they can be a useful indicator of the general state of affairs. Sixty-four households are listed, but fifteen had been standing empty for more than two and a half years, and thirty-one were so poor that they were certified as unable to pay by the churchwardens and overseers of the poor. That left only eighteen tax payers (Document 15). The figure of forty-nine occupied households in 1674 compares interestingly with the figure given by Gardner of fifty-four families in 1634 and seventy-one in 1633. Slow protracted economic decline was undoubtedly having its withering effect on the population of Walberswick.

If the empty households are excluded from the hearth tax returns and persons with the same surname are grouped as if they were one family, a figure of just forty-nine families in 1674 is suggested. This may well be too low, as the hearth tax did not encompass all residents, but a figure around fifty is probable. By the end of the seventeenth century when the second church was being demolished, the south aisle was retained because it could take all those parishioners who came to church, 'which seldom exceeded forty'. In 1752, just two years before Gardner was writing, he says there were only 20 'dwelling houses' with '106 souls' remaining.

Shipbuilding had long been an industry at Walberswick. In fact anywhere there was easy access by road to a harbour, estuary or creek, such as Dunwich creek, or Buss Creek near Southwold, boat and shipbuilding could be found. The town derived income in the late-sixteenth century from shipwrights renting space on the greens where they laid out their timbers. They also had to pay for 'breaking ground' and, laying out keels beside the quay. Floods caused problems when timbers floated away; in 1560 there was 'great loss of boards plank, timber and salt.' Decline along the east coast was recognised nationally in the early 1660s when an Act was passed to encourage ship building in: 'Yarmouth, Ipswich, Aldeborough, Dunwich, Walberswick, Woodbridge and Harwich, where many stout ships were yearly built for the coal and other trade.'

By nature shipbuilding provided short-term employment, contract by contract, and there are indications that itinerant labour was imported to work on some of the larger vessels. When, in 1665, Edward Burford and William Chapman were struggling to recover the Town Duties, they tried to get the shipbuilders who occupied some of the town lands to pay their rent, but were unsuccessful. They complained bitterly that it was the 'ship-carpenters' who most hindered their efforts: 'also they have been a means to bring in many, if not most poor men, women and children into this very poor town, which if these times hold may starve.'

It is clear from the churchwardens' accounts that fires were endemic at Walberswick; there is no doubt they contributed to the physical shrinkage of the town. Surrounded by dry heathland soils, fire could spread rapidly through close-knit timber and thatched houses; open hearths, bread ovens and fish-smoking sheds were a continual hazard. Gardner himself witnessed a raging fire on a windy day in April 1749; growing from a simple chimney fire, it destroyed a third of the town: '... the wind blowing very hard at West carried the burning thatch to the Alms-house, distant about ninety yards, setting the thatch thereof in flames some of which flew above one hundred and thirty yards to another cottage, from which the fire communicated itself to several dwelling houses, barns, stables and out-houses, and in its passage burnt two standing green ashes with several inclosures made with pales ... the violence of the wind, which drove some of the burnt stuff miles off to sea; and in all likelyhood, if tiled houses had not put a stop to the fire, the whole town had been laid in ashes as the wind lay.'

By the 1630s, then, Walberswick had endured a long and debilitating economic downturn. It is against this background that we need to view the miller Turrold's petition of 1636–37, an event which precipitated the enclosure crisis at the heart of the *Bloody Marsh* story.

Poverty and decline: chapter notes

George Crabbe, 'The Village', Book 1, lines 39–48. The contract for Walberswick church tower is published in Lewis 1947; for other towers by the same group of masons see: Chitty 1950. Disputes with Dunwich: Comfort 1994 and Gardner 1754. There is an increasing historical interest in the importance of the medieval fishing industry: Dymond & Virgoe 1986; Bailey 1990; Middleton-Stewart 1996. Pirates, 'Dunkirkers' and 'Hollanders' active along the East Anglian coast: State Papers: 23,508; Acts of Privy Council 105,307. The regulation of the butter and cheese trade, quoted by Gardner (pp. 167–8) appears in the church wardens' accounts: SRO, FC185/E1/2, fol. 9. Payments in alms also appear in the CWA, fol. 16r, see also Gardner p. 166. The Fishmongers' contribution to repairing the haven: SRO, HA30: 50/22/6.15. The sale of the great bell in 1585: CWA, fol. 16r. 'Kadage' and the regulation of the herring fishery: CWA, fol. 62; Gardner pp. 19–20, 145–6, 151. The warrant of 1628 appears in: CWA, fols 134–5; Gardner pp. 170–1. Hoskins 1953 was the first local historian to study the significance of the rise in the price of wheat during the 1620s, but see: Thirsk 1967. The petition for relief of the poor at Walberswick appears in: CWA, fols 134–5. The 'Strangers' of Norwich are summarised in: Ayers 1997, 96 and Priestley 1990. For 'aliens' living in Suffolk in 1568 see Suffolk Green Book No. 12 (1909). The 'burnt' and 'waste' properties appear in the rental of 1635: SRO, HA30/50/22/10.9. Communicants at Walberswick: Gardner p. 160. Migration to New England and the Caribbean see: Tyacke 1951; Lucas 1899, 138–75; R. Warner 1933; Rix 1852. Social responses to dearth see: Walter 1985. Hearth Tax Returns see: Hervey 1905. Number of houses in 1752 see: Gardner pp. 162, 176. Recovery of Town Duties in 1665: CWA, fol. 164; Gardner pp. 180–1. The great fire of 1749 see: Gardner p. 176.

CHAPTER THREE

Turrould's petition

Ill fares the land, to hastening ills a prey,
Where wealth accumulates, and men decay:
Princes and lords may flourish, or may fade;
A breath can make them, as a breath has made;
But a bold peasantry, their county's pride,
When once destroyed, can never be supplied.

Goldsmith

The first petition to the king, prepared in December 1636, came from William Turrould, the miller, who claimed that he was supported by 300 tenants of the manor of Walberswick. According to Gardner there were just seventy-one families three years earlier, so Turrould was claiming that, most, if not all, the population of Walberswick supported his petition, which is most unlikely. Perhaps his supporters included one or two absentee landowners and perhaps some residents of Blythburgh as well. However, it will become clear that although Turrould's cause was a popular one, he himself had mixed support and that Sir Robert Brooke's leasehold farmers, living in new farms on recently enclosed ground, certainly opposed him. Turrould's perception of who was, and who was not, a 'tenant' and hence a member of his community underlies this notion of 300 supporters. However, he listed six 'greate wrongs' and two consequential grievances in his petition (Document 4).

He claimed, firstly, that Sir Robert Brooke was exacting 'excessive' fines on the admission of tenants to their copyholds – ten times more than they were used to paying. Secondly, he claimed that Brooke was preventing the tenants from gaining access to their common pasture for cattle which they had enjoyed 'time out of mind'. Thirdly, he was preventing access to certain of their marshes which rightfully belonged to them. Fourthly, 'Sir Robert Brooke hath caused a great farme house to be built', and, fifthly, he had enclosed a great part of their marsh grounds and commons and attached them to the same farm house, and, 'therein placed tennants of whom he receiveth much rent to the disinherison of your Majesties poore Subjects.' The sixth, and perhaps the most important point, was that Sir Robert had taken away the town quay: 'the benefit whereof in former times was imployed for the repair of the church and releife of the poore, But now the said Sr Robert taketh all the benefit thereof to himselfe, soe that your poor petitioners by the incloseing of the common and marsh grounds and by the losse of the said key are wronged and dammised £200 per annum' (Document 4). Sir Robert Brooke had leased the quay to Henry Richardson, together with all his rights to: 'customs and wharfages of all vessels and tolls on goods coming to or from the key', at Walberswick. When Henry Coke and others had refused to pay tolls on their carts and horses coming from the quay, Sir Robert had begun legal proceedings against them in the court of Exchequer in May 1637.

The consequences were twofold. The remainder of the common was now so overcharged with beasts, cattle and rabbits that the animals were breaking into the townsfolks' own fields: 'and eat up and destroy their grass and corne and what they sow and plant they cannot reape and enjoy'. The burden of these changes and misfortunes was causing the tenants to sell their land 'at a great undervalue' and to 'leave the country'. The use of the word 'country' in this context should be interpreted as 'region' or 'district'; elsewhere it clearly refers to the forty or so parishes that made up the hundred of Blything. However, it is possible, as already suggested, that some were leaving for plantations in the New World. But a more general and poignant complaint was added to the petition: 'by reason of their extreame povertie and the greatness and perversnes of the said Sir Robert are of themselves unable to redress those intollerable wrongs by the cause of your Majesties lawes.'

Turrould first sought to have the matter referred to a committee of nine listed local JPs, including six knights and three esquires: Sir Thomas Glemham of Little Glemham Hall, Sir William Springe from Pakenham, Sir Roger North from Benacre Hall, Henry North Esquire from Laxfield, Sir Edmund Bacon of Redgrave Hall Sir Butts Bacon of Friston (a Norfolk family), Sir Robert Coke and Henry Coke Esquire (also a Norfolk family but with estates at Thorington), and John Scrivener Esquire from Sibton Abbey. With the exception of the latter, they were mostly well known puritan gentry from old established families in the county; unlike Sir Robert Brooke, the son of a London alderman who had purchased his estates just over a generation earlier. Turrould was therefore appealing to the more conservative elements of the county community.

Turrould wanted them, 'or to any four or more of them who are all acquainted with these aforesaid oppressions and that they may have full power to call before them and examine any persons whatsoever for the discovery of the premises and make a final end thereof if they can restore every man his right ...' While the county gentry may have been well acquainted with this dispute, Turrould's statement was one of naive confidence, for the Brookes were well experienced in litigation and although newcomers to the Suffolk gentry had successfully infiltrated the county elite and were soon to marry into it; also their tactics were subtle and not immediately obvious to their rustic but gallant tenants. Furthermore, many of the older county gentry were all too familiar with the problems of enclosure on their own estates and did not necessarily come to the problem with clean hands. The first petition was heard at the court at Greenwich on 23 January 1637. Sir Sidney Montague replied: 'His majesty is pleased to referr this peticon to the committies desired or to any four or more of them to the end they may acquaint the said Sir Robert Brook therewith and require his presence and meeting with them at such dayes and places as they shall appoint' (Document 4).

This message must have been greeted with elation in the streets of Walberswick. But, if the tenants had every intention of seeking an early solution to their problems, Sir Robert Brooke had other ideas. Being a relative newcomer to the area, he may have been apprehensive about having his case heard by his peers; also he was then locked in litigation with the Cokes not only over the quay, but also over evidence of title to his estates at Hinton. Furthermore, in 1637, Elizabeth, his eldest daughter, was coming up to marriageable age: she was in fact to marry Thomas Bacon of Friston, son of one of the JPs in Turrould's petition. So it is understandable that Sir Robert employed delaying tactics, during which time he searched diligently through the court rolls and early charters of his manors for any evidence that would support his case: his notes and those of the lawyers he employed survive among the Cockfield Hall papers. Here he had a significant advantage, for the Hoptons had generated a colossal manorial archive, none of which was available to his opponents. By sharpening the definition of his case he could throw his opponents into confusion by asking questions to which only he and his lawyers knew the answer.

Sometime before May in the following year of 1638, Turrould, exasperated by Brooke's intransigence before the committee of JPs, rather impatiently and perhaps unwisely, returned with his petition to Whitehall: 'the said committees have taken an exact examinacion of the whole estate of the business, they have also mediated with the said Sir Robert Brooke for a peaceable end, to which the committee have found your peticioners very inclyneable yet in respect of the said Sir Robert there labour and indeavors hath been but in vayne' (Document 8).

In 1638 there were other more serious matters, such as the controversy over Ship Money, to keep the king's advisers busy at Whitehall. This was the period of 'Personal Rule', when the king was governing without Parliament and the Scots were raising an army in the north. Ever confident in the rightness of his cause, William Turrould, our intrepid Walberswick miller, appealed to the king's two top advisers in the Privy Council, William Laud, Archbishop of Canterbury and Thomas Coventry the Lord Keeper. He also included the Bishop of Norwich, then Richard Mountagu, who had some local knowledge and in whose diocese Walberswick lay. The matter came before Edward Powell, the secretary at Whitehall, on 2 May 1638 and a week later the Archbishop and Lord Keeper appointed Wednesday 6 June for a hearing (Document 8).

Sir Robert Brooke and Elizabeth his second wife were well connected in London. They also had a house at Langley in Hertfordshire, a more fashionable county, which enabled him to keep in touch with his second wife's family and to visit his mother who lived in the City. There is historical debate about the question of 'localism' in the county community, the degree to which the gentry were more concerned with the affairs of their own county than with that of central government. Brooke, as a representative of the new gentry, clearly had interests in the city and enjoyed a wider social circle than some of the Suffolk county elite to which he was also wedded. It is likely therefore that he was more politically aware than his opponents – aware at least of the difficulties that the king's advisers were having in running an increasingly ungovernable country.

When the 6th of June came, both parties attended from the Wednesday to the Friday: 'with their counsell to their great charges and expenses', but the matter could not be heard. Once more the petition was submitted and their lordships appointed Wednesday 17 October for another hearing. Yet again both parties attended but could not be heard. By now it was autumn and there was little chance of having a hearing before winter. However, in the following spring Turrould once more presented his petition, this time at the Inner Star Chamber on 3 May 1639, where the clerk, Edward Nicholas, appointed 10 May. The case was not heard by the Archbishop and the Lord Keeper until 22 May, but even then it was only partly resolved and rapidly brushed aside (Documents 9 and 10).

Their lordships decided that because the matter concerning right of commonage was something that was determinable by law, the inhabitants of Walberswick should bring a civil action against Sir Robert Brooke at the next assizes, which judgement they declared should be final. It was not unusual for civil actions of this type to be heard at the assizes during the sixteenth and seventeenth centuries. Regarding the quay, Sir Robert claimed this was a matter then being decided by the court of Exchequer and should therefore not be considered by their lordships. However, it was then and there agreed on all sides that it belonged to the town. Much later, John Barwick claimed that it was entries in the town books which had influenced their lordships' decision concerning the quay and this then prompted the townsfolk to look further into the books for evidence about their commons (Document 13). Then Sir Robert claimed that he had spent money on repairs to the quay; so their lordships appointed a commission of local JPs to establish from witnesses how much had been spent and how much had been received in rent by Sir Robert and to report back to the Star Chamber. Thus the matter was dealt with, but it was far from being fully resolved. The next round would take place at the county assizes where such a *cause célèbre* would be bound to raise the temperature of debate.

Once again, if the case was placed before the county assizes, Sir Robert Brooke would be exposed to the criticism of his fellow gentry who regularly assembled at the assizes where JPs were commissioned. One month later – and it is hard to imagine that he was not fully acquainted with the decision when it was made on 22 May – he himself petitioned William Laud and Thomas, Lord Coventry begging for more time: 'your petitioner is most willing to obey the sayd order: but in regard he was not served with nor saw the sayd order untill the 24th of this instant June. At which tyme he was served with the same at his house in Suffolk neare 80 myles from London his Scollicitor being then in London And for that the tyme is now so short and your petitioner so unprepared that he cannot be ready for a tryall at this assizes' (Document 11).

He wanted to put the case off to the following spring, to the Lent term of the assizes. But their lordships must have seen through this ruse, for it was not granted. An action brought by the townsfolk in the name of Edward Howlett was heard in the Summer assizes of 1639.

Turrould's petition: chapter notes

The final lines from Oliver Goldsmith's peom 'Sweet Auburn', written in the 1770s reflect on the 'tyrant's hand' in causing the desertion of his childhood village. The first petition, Document 4, appears among the Cockfield Hall papers: SRO, HA30: 50/22/3.1 [1]. Reference to the disputed quay in the court of Exchequer: SRO, HA30: 50/22/27/3.2. For The Suffolk county community: Everitt 1960. The decision to return to Whitehall with a second petition may also indicate a frustration with the ability of local JP's to settle the matter; see Document 8. The inability of government to come to grips with local issues is characteristic of the period; see Underdown 1985; Holmes 1997. The church wardens' accounts or 'town books' were produced as evidence in court on a number of occasions as stated on the inside cover and elsewhere: cf. CWA, fol. 148. For civil actions at the assizes see: Cockburn 1972.

CHAPTER FOUR

Land, labour and enclosure

Yon is our quay! those smaller hoys from town,
Its various wares, for country-use, bring down;
Those laden Waggons, in return, impart
The country-produce to the city mart;
Hark! to the clamour in that miry road,
Bounded and narrow'd by yon vessels' load;
The Lumbering wealth she empties round the place,
Package, and parcel, hogshead, chest, and case:
While the loud seamen and the angry hind,
Mingling in business, bellow to the wind.

Crabbe

In many respects, Walberswick was an unusual place for seventeenth-century
Suffolk, because it was not primarily an agricultural community. The rental
of 1582/83 lists many small gardens, orchards, canabaria (hemplands) grouped
around the tenements, messuages and cottages in the town, but these seldom
exceeded more than a rod or a rod and a half of land and many were of half
a rod or less (1 rod = 30.25 square yards). Larger crofts and closes lay behind
the cottages some as large as three acres, but most of them were an acre or
less (Map 2).

   Three-quarters of the land in the parish was sandy heath, of very little
agricultural value, and, until the canalisation of the river in the eighteenth
century, the other quarter was barren salt marsh, or mud-flats exposed at low
tide. Between heath and marsh there were a few patches of meadow grazing,
but away from the town the heathland sloped down abruptly into the marsh
making a rapid transition from sandy hill to flat reed-swamp and salt-marsh
(Map 3). Fishing and coastal trade were the mainstay of this small sea-side
community. The transformation of Walberswick from a flourishing town in
the fifteenth century to an impoverished and depopulated village in the early
eighteenth is typical of other small coastal communities many of which also
had to contend with coastal erosion during this period. Yet the pressures which
afflicted the agricultural economy of Walberswick in the seventeenth century,
and which came to bear on the town with such unrelenting force, were similar
to those suffered by many other inland communities which survived relatively
unscathed.

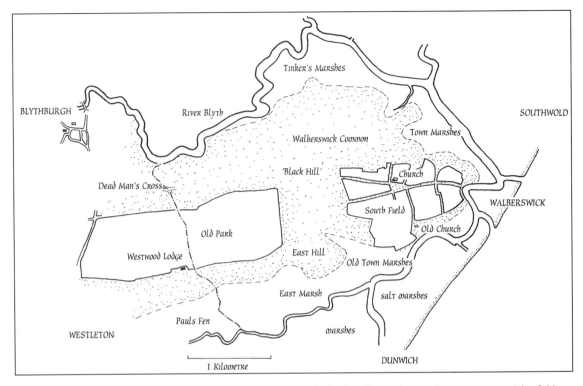

Within the map:

BLYTHBURGH

Tinker's Marshes

River Blyth

SOUTHWOLD

Walberswick Common

Town Marshes

'Black Hill'

Church

Dead Man's Cross

WALBERSWICK

South Field

Old Church

Old Park

East Hill

Westwood Lodge

Old Town Marshes

East Marsh

salt marshes

WESTLETON

Pauls Fen

marshes

1 Kilometre

DUNWICH

MAP 3. The fields, heath and marsh of Walberswick.

Some economic changes hit Walberswick particularly hard, such as the national decline in the fishing industry and the inability of Walberswick's harbour to accommodate the ever increasing draught and tonnage of ships. There was a knock-on effect in crafts related to fishing and shipping trades. Some, such as fish smoking and net making, were by-industries of the fishermen themselves, but others, such as barrel coopering and basket weaving for the packing and transportation of fish were more specialised and may have been supplied from outside. At Dunwich there were gaugers, packers and searchers who were paid to check the packing of herring in thirty-two gallon barrels, while smoked herring were packed with straw in willow baskets tied up with small rope-yarn.

Salt production was a key supportive industry since it was used in large quantities for the preservation of fish. A 'salt house' in the occupation of William Harborne, gentleman, stood by the 'Flattes' near Walberswick quay in 1582/83 (Map 2). Thomas Barweck, a fishmonger of Ipswich, and his father had shares in salt houses at Walberswick which were sold in 1623. Herring was of course bought and sold at the quayside, just as it is today and large quantities of fish were transported inland. Some fishmongers that used the port were not local men and had to pay a toll: 'that any countryman that comith to buy hearinge of any fisherman except of the townes men shall pay to the towne for every 1000 (herring) – 4*d*.'

There were traders in butter and cheese, bacon, corn, timber and coals. In 1583 colliers regularly visited Walberswick: 'that every shipp that comith from

FIGURE 18. The old
Bell Inn and
fishermen's cottages
near the green at
Walberswick. In the
sixteenth century tar
and salt houses
belonging to
Alexander Richardson
and John Barwick's
father stood near
this site.

PETER WARNER

Newcastle being no townsman here shall pay for every that he selleth and is carried way ... 1*d*.'. Salt, coals and 26 barrels of oil, possibly whale oil, were landed by John Barwick, the brother of Thomas above, in 1636. The supply of tar imported from Stockholm was also essential for the protection of ships' hulls and hawsers. In 1582/83 a 'tarhouse' standing on the quay was owned by Alexander Richardson, one of the wealthiest inhabitants (Map 2). Great quantities of hemp, grown on clayland farms and processed as a rural by-industry, were needed for rope and sail cloth. Small-scale hemp production is testified at Walberswick in 1582/83 by the name of 'hempitlane' and the tiny 'canabarea' attached to almost every messuage. And of course there must have been long-distance carriers of fish to inland markets and overseas, as there were at Yarmouth in 1724.

All these trades felt the impact of a declining fishing industry. Yet, although in difficulties in the late sixteenth and seventeenth century, the fishing industry never died out. Also, large vessels were still being built on or near the quay generating temporary employment. Even so, as the fishing and related trades declined so the community became more dependent on agriculture; what is surprising is the almost complete inability of agriculture to sustain them. Most of the wills from early seventeeth century Walberswick clearly indicate that those who were primarily involved in other trades also held small pieces of agricultural land in the parish. John Chettleburgh, shipwright, had a crop of barley growing on the copyhold tenement he had recently acquired in 1625; John Flowerdew, mercer, held freehold lands and tenements in Walberswick in 1619; so did George Borrell, a stonemason. The presence of a stonemason is itself interesting, but one only has to look around a few country houses and at the tombs of the gentry in Suffolk churches to realise that there was more than enough work for a stonemason at Walberswick. We know that the tombstone of John Brooke arrived at Walberswick quay in 1655. On the quayside were landed most of the heavy goods and raw materials needed to supply sixty or so parishes and a dozen country houses.

If Walberswick was no ordinary place, it was certainly no ordinary community. It comprised an unusual diversity of trades and occupations, far more that would be encountered in a small country town or village. Visiting ships brought it into contact with that extraordinary cosmopolitan world of seafaring men. Merchants from Ipswich and London came regularly to Walberswick, to ensure the supply of herring, timber, butter, cheese and wool for delivery to urban markets through the regular network of coastal trade. Every year a few vessels sailed out of Walberswick harbour to join up with the great North Sea fishing fleet which headed up to Iceland. There were casual visitors and Huguenot refugees, customs men and sailors from far and near. The gossip in Walberswick's alehouses was not that of an introspective rural community, but of an outward-looking, diverse crowd, talking about distant and exotic places. Walberswick was not a community of rustics, or even urban artisans, it had the social heterogeneity of a sea port in miniature, it was a cosmopolitan community, semi-urban in character, proud and independent by nature. Such were the folk who opposed the enclosure of their commons.

It is invidious therefore to talk about a 'peasant' community at Walberswick; doubtless there were some small peasant farmers attempting to subsist on agriculture alone, but they were probably in the minority. The majority called themselves, 'mariner', or 'carpenter' as well as 'shipwright', but there were others who were fishmongers, mercers, cordwainers, masons and merchants. Because of their dependence on the sea, and because of the arid heathland soils which comprised three-quarters of the parish, these inhabitants had either very small land holdings, or no land at all. In some cases the agricultural land was held by non-residents who owned larger farms inland. For example, Robert Scolding of Sibton, a yeoman farmer, held substantial farm land at Bramfield and Sibton as well as lands and tenements at Walberswick; similarly Robert Hoxton of Beccles held farms at Sotherton and Henham as well as the free and copyhold tenements, meadows and marshes which he had recently purchased in Walberswick in 1621.

The open fields were confined to a relatively small area around the settlement; they can be determined from map evidence, particularly where there were surviving bits of glebe land around the first church, and from the sixteenth- and seventeenth-century rentals. In 1582/83 the 'Southfield' was made up of 180 small pieces of land grouped into four inequal precincts; it was Walberswick's only open field of any size (Map 3). This area and the enclosed town marshes also close to the town on the north and south sides, represent no more than a quarter of the whole parish. The rest was open commons, saltmarsh, heathland or enclosed sheep-walks and parkland in the ownership of the lord of the manor and controlled by his tenant farmers.

The practice of enclosing heathland into sheep-walk is attested in some of the early maps of the Sandling. For example at Bromeswell and Sutton, parishes near Woodbridge, there is a note on the maps by William Hayward and John Norden marking the ditch of new enclosures dated 1600. Here too there was a dispute over the parish boundary. Farm leases of the early nineteenth century from the same area, describe in detail the banks which separated the fieldwalk boundaries. At Ferry Farm, Sutton, the banks were about eight feet wide at the base with a 'table' on the top where hedges and trees were allowed to grow. The walk banks with their thorn shrub hedges still survive in undisturbed areas of the Suffolk Sandling, including Walberswick, Blythburgh and Westleton.

In the early sixteenth century, before the building of protective sea walls and enclosure of the marsh and the upland sheep-walks, there was remarkably

FIGURE 19.
Sheepwalk boundary surviving at Newdelight Walks on the Westleton–Blythburgh parish boundary. Thorn bushes top a low bank with ditches either side – last vestiges of the sixteenth-century enclosure movement.

little demesne land in the parish. Walberswick had always been a hamlet of Blythburgh and the manor house of Westwood Lodge lay just beyond the parish boundary at the opposite end of the parish from the settlement (Map 1). About half of the once magnificent park of Westwood lay in Walberswick parish and the other half in Blythburgh. There appear to have been no other demesnes in Walberswick. The park was broken up in 1602 and ploughed by Francis Colbie, one of Robert Brooke's new leasehold farmers. Reyce commented in 1618 that: 'this country [Suffolk] cannot glory of so many parkes as it was wont to have, the necessities of this latter age hath given such a downfall of the pleasures of this kind ...' The enclosure of heathland sheep walks in Walberswick as part of the demesne soon followed. It is possible that there was some enclosure of heath ground adjoining the park before this date, there is a mention of 'New Close' associated with sheep-pasture in the Hopton lease of 1534, but this probably lay in Blythburgh (Document 1).

The agricultural resources available to the smaller landowners and tenants in Walberswick were therefore strictly limited. The only access to grazing beyond their limited landholdings were the commons. The economic significance of the common land therefore became disproportionately great for those who had lost their livelihoods in other crafts and industries. In the seventeenth century their landlord was doing the same as landlords elsewhere, he was enclosing the waste, converting it to sheep-pasture and leasing it out to his farmers. The fight to resist enclosure at Walberswick had an added importance for many of the tenants; for them it was a matter of survival.

In 1637 the tenants and inhabitants were adamant that they had the right to graze all their: 'great beasts [cattle] which they can keep on their owne

FIGURE 20.
A felled pollarded oak from Westwood Park resting on the park bank, the victim of modern agriculture. A few pollarded oaks can still be seen in the area of the park, the last remnants of the sixteenth-century landscape.

FIGURE 21.
The old road from Blythburgh, close to Deadman's Cross, site of a suicide's grave, where the road crossed the processional way and parish boundary into Walberswick. The spot is famous for being haunted.

36

lands … upon the great Heath or Common called Walberswick Common at all tymes in the yeare, And that the said Heath extendeth from the townes end of Walberswick to Deadmans crosse', in other words to the parish boundary with Blythburgh (Document 6; Map 1). There are many entries in the church-wardens' accounts detailing 'presentments' for 'great beasts', whereby small sums were paid for the right to graze horses, cattle and geese on the commons. In the late sixteenth century there were problems with incomers grazing their animals without permission: 'If any carter shall bayte his horses upon the common: having lawful warning to the contrary shall pay for every tyme, *4d.*' – a statement which was followed by a mass of animated signatures and personal marks.

FIGURE 22. Typical modern heathland vegetation with hair grasses, heather, lichen, bilberry and gorse. In the past, when the heathland was heavily grazed by sheep and rabbits, a short 'velvety' herbage developed that was largely bare of trees and scrub.

Little work has been done on heathland communities, but in south-east Dorset and the New Forest area of Hampshire Bettey suggests that mixed economies dependent on quarrying, turf, peat and furze cutting, were necessary for even a small population to survive. There is a distant history of peat digging in the Blyth valley and there is a 'Turfpit Marsh' on the tithe map for Westleton at the western end of Westwood marshes, but there is no reference to peat digging as part of the dispute over commons at Walberswick in the seventeenth century. However, the marshes and estuaries were rich in all manner of bird life, just as they are now. Reyce lists the 'sea fowle' that might be expected, and it is implied might be eaten: 'those we call seapies, coots, pewits, curlews, teal, wiggeon, brents, duck, mallard, wild goose, heron, crane, and barnacle [goose]'.

Most abundant were the rabbits or 'conies' that overran the heath, nibbling down the grass and vegetation to a fine velvet. The livestock animals that grazed on heathland pastures were notoriously under-nourished small and hardy creatures. Some communities made a living from breeding these animals, such as the famous small ponies of the New Forest called 'heath-croppers'. Reyce also mentions horses in seventeenth-century Suffolk bred on upland stony ground as being: 'most puissant, and strong for service, of quick life, and spirit, of high pride, and most comely shape'. It was these rugged, scruffy creatures that Sir Robert Brooke seemed to find most objectionable grazing below the gardens of Westwood Lodge on Pauls Fen.

The upland commons at Walberswick provided desperately thin grazing, but the extent of heathland, as much as 800–1,000 acres in the sixteenth century, allowed for large flocks of sheep. In 1637 the townsfolk claimed that their flock numbered between three and four hundred (Document 6). There were two fold-courses belonging to the lordship. The demesne flock, sold by Arthur Hopton to Sir Edmund Rouse in 1541, probably contained about another six or seven hundred sheep, but they grazed into Blythburgh as well as Walberswick. In July 1646, after the death of Sir Robert Brooke, John, his son, purchased the entire demesne flock back from Thomas Palmer, his leasehold farmer, for the large sum of £426. It was said that there were then 726 sheep, including wethers, rams, ewes and lambs.

The tenants' flock grazed over the marshes, open fields and commons on the eastern side of the parish adjoining the town, and was usually attended by a shepherd (Documents 5 and 6); while the demesne flock grazed over the park, marshes, commons and enclosed ground extending across the parish boundary into Blythburgh. Such was the great tract of open space that separated these two areas that there was no need for a hard and fast boundary between them. Sir Robert Brooke claimed that the heathland contained just 500 acres (Document 5), but John Barwick claimed the heathland commons extended to 1400 acres (Document 13). Such figures were exaggerated or understated by either side to suit their argument; according to White's trade directory the entire parish in 1844 consisted of 1771 acres and that included 130 acres of saltmarsh and commons, but even this figure may be no more than an estimate.

It is impossible therefore to arrive at any sensible figures for commonland acreages from the seventeenth-century legal documents, firstly, because neither side can be believed and secondly, because the position was clearly changing all the time with enclosure, temporary cultivation, and reversion to heath again.

When the tenants' sheep strayed into Sir Robert Brooke's sheepwalks he fined them a penny for every stray animal. The towns' great beasts could graze all over the commons, but they had to be attended by a 'follower', usually a child, to prevent the beasts trespassing into the walks, or any of the ground periodically sown with corn; if they strayed they were impounded (Document 5). The lack of any clearly defined boundary between areas of common grazing was one of the principal difficulties in trying to resolve disputes over enclosure. For example, the Brookes claimed that the marshes below Westwood Lodge were part of their demesne, while the town insisted that they were grazed as part of their commons (Documents 5 and 6). The Brookes also claimed that the townsfolk could only graze their sheep as far as Black Hill, about a quarter of a mile out from the settlement, but few people seemed to know exactly where Black Hill lay.

A further complication arose from the periodic cultivation of the heath. This was undertaken by the lord of the manor or his under-tenants, the crop was usually barley, but a 'great part' of it was reaped and carried away by the commoners, presumably in acknowledgement of the grazing rights that they had thereby lost (Document 6). Once enclosed as sheepwalk and leased out to Sir Robert's farmers the ground available for this type of periodic cultivation was greatly diminished. Thus, the townsfolk began to lose income from the heathland which had supplemented their very small land-holdings.

The rabbits which abounded on the sandy heaths were a regular source of food and income to both the lord of the manor, who had a great coney warren in his park at Westwood, and his tenants: 'divers of the inhabitants did frequently in the day time with doggs and Netts take connyes wich did burrow and breed on the heath' (Document 6). The sheep and rabbits together reduced the grasslands to a 'short velvety herbage of grass and moss' which was 'studded more or less with furze or whin bushes'. In other areas ling or heather blanketed the heath, but even here the sheep and rabbits could find sustenance, in bilberry and lichen. There are still three or four different species of lichen that grow on the heath but *cladonia impexa* probably provided most of the grazing for sheep among the heather.

By the mid-seventeenth century, the upland commons which had once given the townsfolk access to large areas of rough grazing, rabbitting and the possibility of an occasional share in demesne crops, however meagre, were now significantly reduced. It might not have mattered if their main livelihood of fishing had prospered, but the enclosures came at a time when the community was turning towards agriculture for subsistence and survival. The enclosures and accompanying threats from their landlord and his farmers were therefore seen as a body-blow to the community itself. In reality their position was dire. The soil was so poor that it is hard to see how the town could have subsisted on agriculture alone even if it had had access to all the land in the parish. However, to many of the inhabitants it must have seemed as if the preservation of these commons and the crops grown on them, made the difference between survival and economic extinction.

What drove this relentless desire for enclosure? Was it a matter of simple economics, or, as some might argue, the greed that comes from the prospect of making very large sums of money? Certainly the Brookes were agricultural improvers who sought to increase crop yields and extend the pastures under their direct control, and to establish new farms that could be leased for higher rents. The intensification of agriculture at this period was driven partly by the falling value of rents, caused by inflation, and partly by the rising price of grain. During the period between the mid-fifteenth and the mid-seventeenth century the price of barley saw an eightfold increase. Barley, and to a lesser extent, rye, was ideally suited to the light Sandling soils of the Suffolk coast. It was said that the best barley for brewing should have a taste of salt in it, in other words it should be grown within sight and sound of the sea. Whatever the truth of this, barley along the Suffolk coast commanded a good price. But the heathland soils could not sustain regular cropping. They were desperately dry podsols, extremely acid and leached of all nutrient. For hundreds of years they had been grazed as uplands and only used for periodic cultivation with long periods of fallow, sometimes up to ten years between ploughings. Cattle and sheep that grazed the heaths as summer pastures were selectively culled during the winter. The result of this process was a nitrogen cycle which drained the heathland pastures of nutrients.

To improve yields on the Sandling, three things were needed. Firstly regular and heavy manuring: regular to restore the nitrogen cycle and return goodness to the soil, and heavy because the soils were quick-draining and there was a tendency for the goodness to leach out. By grazing sheep on the richer meadow pastures and folding them at night on the upland heaths their manure could be concentrated on selected areas. Secondly, the heaths needed marl, calcareous material such as chalky-boulder clay to improve the consistency of the soil and reduce the extreme acidity of the heathland podsols. Marling has a long history in this area, at least one marl pit is mentioned near Thorington church in the early thirteenth century. Thirdly, they needed rain – the most difficult thing to supply it might seem. However, in the period between 1550 and 1850 there was what is known as the 'Little Ice Age', with heavier rainfall and colder winters. In the years between 1550 and 1680 it is estimated that about 20% of the annual rainfall fell between July and August. This was a serious disadvantage for many farmers, but for those on lighter soils it was a god-send.

There were two more ingredients, deep ploughing and access to rich pastures for the sheep to graze during the day. The nearest pastures were the meadows and salt marshes below the heathland commons, but there were not enough of them and they were subject to saltwater tides. So to make more meadows, drainage and flood-walls were needed to keep out the tides. With more meadows, more sheepwalks could be created and so more of the upland could be ploughed to grow valuable barley crops and rye. These valuable cereal crops were the prize, the sheep were also valuable, but cereals were the economic driving force behind agricultural improvement at this period. To build flood walls and drain the meadows, capital was needed, the Brookes had the capital and the aggressive determination to carry such projects through. There can be no doubt that their efforts at Westwood Lodge were successful; by the late eighteenth century the farm was described by Arthur Young as: 'without exception the finest farm in the county.'

Before the eighteenth century when the river was embanked and 'canalised', nearly all of the marsh was subject to saltwater tides. Only salt resistant vegetation, such as sea fern-grass *(Catapodium marinum)* and red fescue *(festuca rubra)* would have offered some limited grazing for sheep. There may have been an attempt to build tidal banks in the Middle Ages, but the first clear indication of embanking and enclosure comes in the late sixteenth century at the instigation of Alderman Brooke. In 1637, Pauls Fen and East Marsh were said to have been: 'inclosed by a wall from an arme of the sea about 40 or 50 years since', and that it cost between one and two hundred pounds – the figure may well be exaggerated (Document 5). There had been a shallow open water broad here up to this time. In the seventeenth century, as a direct result of drainage the area became valuable marshland grazing, it changed also in

Bloody Marsh

FIGURE 23. Tinker's Marshes on the north side of Walberswick, looking north towards Bulcamp and Henham. The group of marshes closest to Tinker's Farm may have been embanked and drained as early as the sixteenth century. Others were drained following canalisation of the river in the eighteenth century.

terms of ownership: 'this our common fenn which they have since called Est Mearch.' (Document 13) The group of four marshes north of Tinker's House may also date to about this period; they must have been taken out of open saltings, the sort of landscape to which they are now reverting.

The Brookes were following a national trend. They were grocers from the city of London and knew full well the steep rise in the price of agricultural produce that had taken place in the late sixteenth century. The upgrading of saltmarsh into meadow pasture required massive capital investment and the Brookes had the money to invest. They also understood the long-term economic implications of developing their estates. There were laws against enclosure dating from 1597, but they were not effectively enforced. Had the Brookes been deliberately depopulating their villages and turning arable land into pasture, there would have been a case against them and they would have risked a stiff fine, but that was not the case. The outcome of banking the marshes was to transform the poorest of wet land into prime grazing. A similar process was taking place in the fenlands of Cambridgeshire in the early seventeenth century. There the landowners combined their capital to create the 'Adventurers' fens. Some co-operative effort of a similar kind, but on a much smaller scale, by the townsmen of Walberswick is implied in their claim that they made the: 'passages and draynes unto and in ... Pauls Fen' (Document 6).

Similar marsh-land enclosures can be seen in other East Anglian estuaries, such as the Deben and Alde where there was easy access to London via the coastal trade. At Bromeswell, the Town Land was an enclosure made in the marshes before 1601. The sheep that grazed these marshes could be penned at night on the dry upland pastures where their manure could prepare the heath for regular cultivation. It is likely that the townsmen of Walberswick were attempting to drain and improve their common marshes in just the same way as the Brookes were improving theirs, but they lacked the capital and the expertise to undertake any large-scale drainage schemes. While they maintained the 'causyes, hanges and passages' for their animals to go into the marshes, flood walls were the work of the squire, although they grumbled that some of the walling, 'might then have byn done with lesse charge' (Document 6). The tenants of Walberswick lived in a rapidly changing world and were as much victims of national economic change as they were victims of a rapacious landlord.

Two other factors under-pinned these changes. Firstly, there was the willingness, even determination, of landlords to force enclosure on their tenants. In some cases this may have been driven by economic need, but that was not the case where the Brookes were concerned. Secondly, it has been estimated that nationally, in the period before 1680, there was a major shift from copyhold to leasehold tenure, caused largely by inflationary pressure. This was by no means uniform, nor was it always achieved by the same means. A landlord could effectively force a customary tenant into a short-term leasehold tenure if the only alternative was paying a punative arbitrary fine to enter an inherited copyhold tenement for life. Thus, by controlling rents, the landowner could keep pace with inflation and share in the rising market prices enjoyed by the farmer. Also the increasing need for capitalisation in farming could result in the engrossing of small farms and putting small landowners out of business, just as it does today.

The sheer weight and range of economic forces during this period worked in the landlords' favour. While the rising price of agricultural produce and the effects of inflation, particularly the relative fall in wages, benefited the larger farms, the peasant farmer, on the other hand, who might be dependent on selling his labour to make ends meet, could be forced to sell up and pay off his debts. So the large farms grew larger as the smaller peasant holdings withered away. For many dispossessed small farmers in the seventeenth century the open road and a life of vagrancy beckoned, Walberswick was the same as anywhere else in this respect. The late 1620s in particular were a time of national dearth when these problems came to a head, resulting in dearth riots in some parts of East Anglia. It is remarkable perhaps that there is no evidence for disturbance as a consequence of food shortages at Walberswick.

By 1628 there was such grinding poverty in Walberswick that eighty poor persons there, it was feared, might perish for want of food. The enclosures were not directly responsible for this poverty, but they may well have put the smaller landowners out of business and ruined some families. Vagrancy was forbidden by the Tudor Poor Law Acts, and paupers were expected to be maintained by the overseers of the poor in each parish who levied a poor rate on other residents. Thus at Walberswick poverty generated poverty, until the whole community was well nigh destitute.

Land, labour and enclosure: chapter notes

George Crabbe's most famous poem 'The Borough' reflects on Aldeburgh in the early nineteenth century: Poster 1986, lines 69–78. Packing fish: Gardner pp. 19–20. Thomas Barweck's will: Allen 1989, p. 485. Tolls paid by fishmongers and colliers: CWA, fol. 10r. The 'tarhouse' of 1582/83: SRO, HA30: 50/22/10, fol. 13. Hemp: Evans 1985. Transportation of Yarmouth herring in 1724 see: Defoe 1991. Wills of John Chettleburgh, John Flowerdew and George Borrell: Allen 1995, p. 312; 1989, p. 36, 202. Wills of Robert Scolding and Robert Hoxton: Allen 1989, pp. 169, 307. Maps of Bromeswell and Sutton: SRO. V5/22/1; JA1/48/2. Lease of Ferry Farm, Sutton 1815: SRO, HB17: 52/15/3. The decline of parks: Harvey 1902, p. 35. Presentments for great beasts: CWA, fol. 38. Fines for feeding horses on the common: CWA, fol. 10r. There was a similar problem in 1658 with incomers overgrazing marshland commons further upstream at Wenhaston causing the 'better orderinge and regulating' of the common: SRO, HB26/371/73. New Forest area: Bettey 1987, pp. 19–22. Wildfowl and horses: Harvey 1902, pp. 42, 47. Sale of sheep in July 1646: SRO, HA30: 50/22/3.21 [8]. Different acerages of common: CWA, fol. 148; Gardner 172; White 1844. Fines on stray sheep: SRO, HA30: 50/22/3.21 [15]. Dispute over Black Hill: SRO, HA30: 50/22/3.1 [55, 57]. Heathland vegetation see: White's 1885 Directory of Suffolk, p. 55. and for lichen see: Armstrong 1975, pp. 80–3. The rising price of barley: Thirsk 1967. Inflation: Outhwaite

Land, labour and enclosure

1969. Marling: Armstrong 1975, pp. 105–6; Warner 1982, p. 5. Climatic change: Lamb 1966. Westwood Lodge Farm in the eighteenth century: Lawrence 1990, pp. 52–7. Change of name from Common Fen to East Marsh: CWA, fol. 148. Depopulation in relation to enclosure see: Beresford 1961; Hoskins 1973, pp. 57–9. Fenland enclosure and drainage: Derby 1940; Lindley 1982; RCHM., 1972, lv–lxii. Bromeswell 'Town Land' marshes: SRO, V3/22/1. The shift from copyhold to leasehold: Spufford 1974, p. 50; Thirsk 1967. Johnson 1909, is the formative study on the disappearance of the small landowner. For dearth riots in East Anglia see Walter & Wrightson 1976.

CHAPTER FIVE

Sir Robert's answer

JUNE: SHEEP

In this month, the flocks of stock sheep are regularly
managed: they live on the commons and sheepwalks,
with little change or trouble. The stock intended for
fatting, such, for instance, as wethers bought in in April
or May, and intended to be sold fat from turnips or
cabbages the following winter, should be kept not like
fat sheep, but throughout this month on the poorest of
your food: they may be turned on to your commons or
sheep walk, or into your bare pastures, and kept so for
eight hours a-day...

Arthur Young (1771)

William Turrould's petition took several years to be heard. It was initiated
before January 1637 and although the matter of ownership of the town quay
was resolved in May 1639, the case concerning right of commonage was not
dealt with until the summer assizes of that same year. Part of the problem
was the intransigence of Sir Robert Brooke and his reluctance to expose himself
to the judgement of his fellow JPs. Yet it is clear from the Cockfield Hall
papers that Sir Robert had carefully prepared his defence of the issues raised
in Turrould's petition, and that these detailed arguments were ready to be
presented at Greenwich as early as 23 June 1637.

Brooke had the advantage of documentary evidence in the form of court
rolls and the Hopton archive that he had purchased with his title to the land.
However, he also took care to find witnesses, particularly old men who could
remember the way the commons and marshes had been managed in the past.
Any references to earlier enclosures were vital because they demonstrated a
precedent that might support his case. One of the underlying issues was the
degree to which his rights as lord of the manor prevailed and the degree to
which the freehold rights he had purchased gave him freedom to do as he
pleased with his land. Conversely, there was the question of the degree to
which customary rights enjoyed by the tenants 'time-out-of-mind' impinged
on the freedom of his lordship: 'He humbly desireth the benefit of a subject
for preservation of his inheritance according to law' (Document 5). Clearly
the freedom of the squire impinged on the freedom of his tenants; this

Sir Robert's answer

contradiction in 'freedoms' was a theme that ran through much of the legal debate.

There is no doubt that the previous lords of the manor at Walberswick, particularly Sir Owen Hopton, had enjoyed an easy paternalistic relationship with their tenants. The estate in its later years under the Hoptons may not have been run very efficiently, but the tenants enjoyed considerable freedom; with currency inflation during the Tudor period their rents had gone down in value while the price of agricultural produce had risen. A tenant in such a position could hardly fail to prosper in the late sixteenth and early seventeenth

FIGURE 24. Ruined windpump near Westwood Marshes looking towards East Hill (*left*) and Hoist Wood (*right*) in the distance. Thomas Gardner mentions a 'mill' and sluice being erected on this site in 1743. This is the area of 'East Marsh' adjoining the 'Old Town Marshes on the east. In 1644 John Barwick claimed 'this our common fen they have since called Est Marsh: this also by violence hath been kept.' In 1582–83, Robert Hoiste held the land which is now Hoist Wood.

century. Now the Brookes wanted to run the estate more efficiently, by enclosing waste ground and banking and draining the marshes, by building new farm houses and charging realistic rents.

Sir Robert could easily demonstrate that he was the legal owner of the manor of Blythburgh, which included the hamlets of Hinton and Walberswick. He was always careful to stress the subordinate nature of the community at Walberswick, which was technically correct. Walberswick was still no more than a chapelry of Blythburgh and its tithes, which had originally belonged to the Priory at Blythburgh, were now in his hands. The fact that Walberswick had grown in size to rival Blythburgh, and the priory had been dissolved a hundred years before, was irrelevant. This stress on ancient legal precedent is crucial to an understanding of the dispute. Brooke was arguing for the 'ancient constitution', while at the same time instigating fundamental changes. His opponent, William Turrould was doing much the same thing: arguing for the retention of ancient customary rights, but at the same time, because of economic pressures, making much heavier use of them. While Brooke sought evidence for past practice in his court rolls, Turrould researched the town books for evidence to the contrary. Neither party was arguing for revolutionary change, they both wanted 'ancient rights', but on fundamentally different terms. Such inherent conservatism may well be 'normal' in a stable society, but there were fundamental and irresistible economic and social changes causing friction in seventeenth-century society, fanatical religious beliefs created tensions on both sides, and together these mixed to form a potentially explosive political cocktail.

Brooke was careful to gainsay the accusation that he was charging excessive fines, 'which they pretend to be ten tymes soe much as have been usually paid.' He argued that the fines were, 'arbitrable at the will of the Lord' and that he had never taken more than: 'two yeares proffitte of any one tenament for a ffyne according to the full value.' (Document 5) This was a careful way of saying that he was now charging realistic rents and that the entry fines were a fixed multiple, about two years' worth of the true rent value. His tenants may not have been happy with having to pay realistic rents, but such is life, he could not be expected to subsidise them. With inflation in the economy it is likely that some tenants had enjoyed declining rents over many years. Faced with renewing a tenancy on the death of a father or grandparent they would be faced with a massive increase in rent and an impossibly large entry fine, but they should be thankful for the good times they had enjoyed, not complaining about the realistic rent they were now having to pay. These were realistic rents which reflected the market value of tenancies, particularly leasehold farms.

Sir Robert denied that he had taken away the tenants' marshes, 'wich they pretend to be their common'. These he said were always part of the demesnes of Westwood Lodge, his manor house in Blythburgh. The marshes which they claimed as common had been enclosed 'with a bank from the sea', which had cost between one and two hundred pounds forty years earlier in about 1596

FIGURE 25. The
burnt-out shell of
Tinker's Farm,
destroyed by fire in
1999. The farm is
mentioned in the
eighteenth century,
but this or its
predecessor may have
been one of the new
tenant farms created
by Sir Robert Brooke
in the seventeenth
century.

(Document 5). Furthermore, Sir Robert had purchased 16 acres in 'Paules fen' from a 'Mr Cannon' shortly after his father's death in 1601, and this abutted onto the marsh of the lord of the manor, also called 'Paules fen'. Furthermore, he had witnesses who could testify that Pauls Fen lay in Blythburgh and was taken in by the inhabitants of Blythburgh when they beat the parish bounds (Document 16b).

There were then, just as there are today, two extensive areas of marsh-land north and south of the town of Walberswick. In 1582/83 the 'common marsh' bordering the river north of the town was known as 'Great Copdale', while 'Blackness Common' and the 'Town Salts' coincide with the 'Old Town Marshes' of today in the area east of Stock's Lane (Map 1). The question was: how far did these marshes extend and how far were the townsfolk allowed to graze their animals into them? But why was this a cause of concern to Sir Robert? It is clear that Pauls Fen, or certain parts of it, lay in the area immediately below the gardens and south front of Westwood Lodge. No doubt the rag-tag and bob-tail of tenants' animals was not a pleasing vista and detracted from the great country house and its otherwise idyllic setting.

Sir Robert flatly denied that he had built a farmhouse on the tenants' common, but only upon his own freehold land which he had legally purchased. Where was the farmhouse in question? The answer is that there were probably at least three, if not four leasehold farmhouses established by the Brookes. Their exact location cannot be identified with certainty. The most likely candidate is the extremely isolated farm called 'Tinker's House' adjoining Tinker's Marshes to the north of the parish. Here the farmhouse has a group of marshes enclosed by a wall that fits the description in the archive (Map 1). Another is Eastwood Lodge Farm on the road between Walberswick and Blythburgh. Another candidate may be lost beneath 'Burnt House Marshes', a name that appears for a group of marshes on the Walberswick tithe map for 1841, south of Sallows Walk. These marshes were the most hotly contested in the dispute and to which the name 'Bloody Marsh' was applied. Turrould's petition is emphatic that the farm house was built on the marshes; Tinkers and 'Burnt House' are the only two sites which fit the description. At the time of writing there is no archaeological evidence to support the idea that there was a house on 'Burnt House Marsh', but the name speaks for itself. Eastwood Lodge Farm is in an upland situation and therefore cannot be the farmhouse complained of by Turrould. Nevertheless, its situation near 'Black Hill' and the supposed limit of the commons makes it a likely candidate for a Brooke establishment.

Concerning the complaint of surcharging the commons, Sir Robert was circumspect in his answer. Firstly, he maintained that the ground which the

FIGURE 26. Straw bales standing on sheepwalks at Eastwood Lodge, Walberswick. This light but productive agricultural land is the legacy of sixteenth- and seventeenth-century enclosure of heathland and years of careful husbandry.

tenants claimed to be commonable was part of his ancient 'charter warren' and 'sheepcourse' belonging to his manor of Blythburgh. Secondly, he did not deny that the tenants had a right to graze, 'their great cattle with a follower when it is not sowen with any kynd of grayne.' Thirdly, he retained the right: 'to feed the same with his flock of sheep replennish and keep the same for a warren or plough up the same at his pleasure.' In addition: 'when the same or such parte thereof is sown with any kind of grayne that then and soe long they ought not to have any feed for their great beasts in any places soe sown' (Document 5).

True, Sir Robert held an ancient charter of free warren dating from the time of Henry II, but this was little more than a licence to hunt game. There had originally been deer in the park of Westwood, but these had been 'destroyed' in 1602, when the park, gardens and outbuildings at Westwood Lodge had been leased. Under the terms of this lease, Francis Colbie the lessee, was allowed to kill all the deer, but had to leave 4,000 conies (rabbits) on the estate when the lease terminated. The rabbits were then valued at 36 shillings per hundred; so their potential value was somewhere in the region of £72 per year. The tenants claimed that they had a right to net rabbits using dogs on the common; clearly this was an area of conflict. The rabbit warren was an important asset to both the Brookes and the leaseholder of the park, well worth protecting from potential poachers.

The sheep course involved the ancient practice of gathering all the lord's and tenants' sheep into one flock, grazing them on the lowlands, marshes or root crops during the day, moving them to the light sandy uplands in the evening and then penning them at night on light land which would subsequently be ploughed and cultivated. Thus the manuring and 'tathing' of the sheep enriched the hungry soils of the Sandling, making periodic cultivation possible. If the sheep were penned on the tenants' land they paid for the goodness that the sheep imparted even if they had some of their own 'cullet' sheep mixed in with the lord's flock. The system was very similar to that practised elsewhere, in Norfolk and on the chalk downlands of southern England.

One of the complications of this dispute was the distinction between 'great beasts', including cattle, horses and sometimes geese, which were allowed to graze the commons, and sheep which could not. In the churchwardens' accounts there are many references to 'Presentments of Great Beasts' where townsmen paid the churchwardens for use of the commons. For example, in 1580: 'Robert Smith dothe present his great beastes that is to say 4 nett bestes [heffers] and one horse beste … 5s.' These small payments made a significant contribution to the parish coffers. There are also many references to Brooke fining his tenants a penny for every sheep that strayed onto his demesnes. Part of Sir Robert's programme of improvements for the estate must have been to do away with not only the mixed flocks of tenants' sheep, but also the problem of animals wandering all over the place. They posed a serious threat to agricultural improvement through the spread of disease and inbreeding – there was little point in going to the great expense of enclosure if it did not result in the control of animals and their selective breeding.

That the heathlands and commons were periodically ploughed was amply testified: 'It appears by the rigges and furrowes on most parte of the heath that the same have usually byn plowed.' Indeed, there are many places on the heath today where the remains of ridge and furrow can still be seen, most notably in Westleton parish at 'Black Hill Heath' close to Walberswick boundary. There was evidence also from the court rolls that five tenants had been fined in the past when their cattle had strayed over the places sown with corn, that was why it was essential for the beasts to have a 'follower' or cowherd. It was stated: 'that the lords of the said manors and their farmers have used to plow such part of the said walk or heath as they would, and when any part thereof was sown with corne the inhabitants of Walberswick did not put their cattle upon any such places soe sowen until the corne was reaped, but if their cattle did stray and come on the corne they were impounded' (Document 5).

Brooke utterly denied the tenants' claims that they were forced to sell and leave the 'country'. Over this he was most adamant, for it was illegal to cause depopulation by enclosure and if this were proven he could have been subject to a substantial fine. 'Wheras they Complaine that diverse of them by this Respondents unjust dealings with them have been constrayned at a great undervalue to sell away their land and to leave the Country: he utterly denyeth that he hath byn the cause that any man sould soe doe' (Document 5).

Sir Robert's answer: chapter notes

Arthur Young, *Farmer's Kalendar* 1771, is typical of the astute professionalism that was beginning to develop in the farming industry one hundred years earlier; no mention is made of others who might be dependent on commons and 'bare pastures'. Document 5: SRO, HA30: 50/22/3.1 [42 and 51] summarises many of the issues in this chapter. For the agricultural background see: Thirsk 1967. Walberswick Tithe Map and Apportionment: SRO, FDA267/A1/1. For rabbits sold to London poulters see Richmond 1981, p. 38; Becker 1935, pp. 63–4. The folding of sheep: Allison 1957; Bettey 1987; Armstrong 1975. Presentment of beasts: CWA, fol. 38 and fines for stray sheep: SRO, HA30: 50/22/3.21 [15].

CHAPTER SIX

God and Mammon

....

Still it occurr'd that, in a luckless time,
He had fail'd to fight with heresy and crime;
It was observed his words were not so strong,
His tones so powerful, his harangues so long,
As in old times – for he would often drop
The lofty look, and of a sudden stop;
When conscience whisper'd, that he once was still,
And let the wicked triumph at their will;
And therefore now, when not a foe was near,
He had no right so valiant to appear.

Crabbe

Most visitors to Walberswick today are drawn to its small well-kept church standing within the ruins of a far larger one; not only is it a constant reminder of Walberswick's forgotten past, of its endurance and survival during good times and bad, but it is still the most significant historic monument in the parish. The church was central to the life of the seventeenth-century town, just as it had been in the Middle Ages. The three churchwardens, elected annually down to 1628, and bi-annually thereafter, together with two overseers of the poor, two questmen, two surveyors and the sexton were the most important parish officers. Together they formed the local government of the community, answerable to the Justices of the Peace, including of course the lord of the manor, who occasionally signed the accounts and the periodic visitation of the Bishop or his representatives. The clergy came and went at the appointment of the lord of the manor and are barely evident in the churchwardens' accounts. The ministers or chaplains served the much larger parish of Blythburgh and sometimes also Southwold, and so had little time to interfere with day-to-day parish business.

Really we should consider three churches at Walberswick as the guide book does, the two we see today, one inside the other, and a third earlier church about a quarter of a mile to the south near the marshes, nothing of which remains above ground. Continuity of worship on one site is to be expected for most medieval parish churches, which often contain evidence for structures dating from the twelfth century if not earlier. Even when the focus of settlement moved elsewhere, the old church was rarely abandoned, although it might

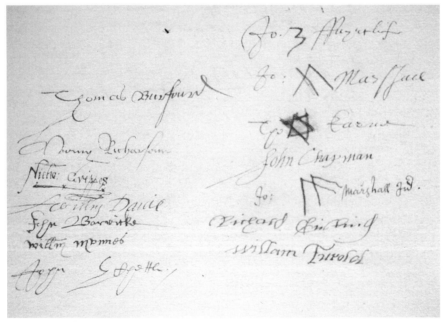

FIGURE 27.
Walberswick church
as it appears today
from the south east.
The road in front of
the church was known
as 'Fishersway' in
1582–83.
PETER WARNER

FIGURE 28.
Signatures of the
Walberswick
churchwardens and
parish officers in the
church wardens'
account book c. 1644.
William Turrould
bottom right.
COURTESY OF SUFFOLK
COUNTY RECORD OFFICE

remain in isolation. There are many local examples, such as Westhall, Brampton and Wrentham where the church stands some distance from the modern focus of settlement. At Walberswick the community decided to make a break with the past sometime in the early fifteenth century when they re-built their church on a much larger scale near the new settlement and left the old church in ruins. This decision was itself indicative of an underlying instability and a lack of continuity evident throughout the town's history.

The site of the first church can be identified from early documentary sources. The rental of 1582/83, copied from late medieval abuttals, gives its location at the end of the 'common way leading towards le old church' and 'le old Churchway leading to Le Key', apparently equating with the modern Stocks Lane (Map 2). It was also marked on Gardner's version of Agas's map of 1587 as 'Walberswick Old church' very close to 'Walberswick Key'.

The earliest reference to a chapel at Walberswick comes in the charters of Blythburgh Priory. On 22 February 1279, the prior and canons came to an agreement with the lord of the manor of Walberswick, then Robert fitz Roger, that they would supply a suitable chaplain to serve in the chapel of Walberswick. This chaplain was to provide all the sacraments of the church for all the men of Walberswick as and when required. The prior would provide a suitable lodging for the chaplain, 'Prystes Lane' and the 'Prystes Close', which had once had a building on it, are described in the 1582/83 rental (Map 2). In return Robert fitz Roger quit-claimed the right of his men and

FIGURE 29.
Blythburgh church floodlit. Although Walberswick was technically a hamlet and dependent chapelry of Blythburgh, by the fifteenth-century Walberswick church was comparable in size and splendour to its parent, Blythburgh.

serfs to partake in the Christmas feast that the monastery had 'long been accustomed to provide' for the men of Walberswick. Robert also benefited from the chaplain saying masses for his soul, but he also had to guarantee the tithes of fish paid by his men during the herring, smelt and shallow water fishing season, which his bailiffs '... shall by distraint compel full payment ...' This was probably an important part of the deal for the prior. Gaining full sacramental rights and a permanent chaplain raised the status of the local community; a fact the parishioners would have appreciated. Walberswick had come of age, but the church was always referred to as the 'Chapel of St Andrew' even as late as 1582/83 and technically it remained a dependent chapelry of Blythburgh. The rate of tithes on fish was then carefully set out and agreed in a later charter of 1411.

Apart from a scatter of stones in the area marked on Agas's map nothing survives of the first church. There are two possible reasons for transferring to the present site. First, the physical change in the location of the harbour which began to take place from 1328 when the river mouth moved about a mile north from Dunwich. Secondly, flooding may have forced a move to higher ground. We know that there was a serious problem with flooding along the coast in 1362, when Leiston Abbey, once in a similar low-lying coastal situation, was flooded and moved to higher ground. However, the movement of the church northwards at Walberswick comes almost sixty years after this period of flooding.

Problems with the new harbour continued into the late fourteenth and early fifteenth century. In 1399, the new harbour was itself blocked, and a cut was dug on the north side of 'Passely Sands'. Agas's map marks six 'key's or 'port's in Walberswick and one in Southwold (Map 2). One lay close to the oldest church, mentioned above, another 'Walberswick Key' ran along the south side of the present river Blyth opposite Blackshore; a third site called the 'Old Key' lay on a bend of the river just south of the present harbour, and a 'New Key' lay on the Southwold side of the river. A site called the 'New Port' occupied a position at the junction of the river Blyth and Dunwich Creek, immediately south of the present village of Walberswick, and opposite the river mouth as illustrated in 1587. When this latter site was subject to coastal erosion in the 1960s late medieval pottery was revealed suggesting that the 'New Port' was new in the period after the river changes of the early fourteenth century.

A determining fact is that the present church with its fine tower occupies high ground well inland and about equi-distant from all of Walberswick's 'ports' and 'quays'. The conclusion therefore is that flooding and the continually changing nature of the harbour mouth caused the church, with its splendid high tower, to be built in a safe and central place, visible not just to everyone in the parish, but also to fishermen far out to sea.

The present tower was commissioned in 1426. The contract between Richard Russell of Dunwich and Adam Powle of Blythburgh, masons, and the churchwardens is a rare survival being one of only four such contracts known

FIGURE 30 (*opposite*). The east window of Walberswick chapel, reconstructed in 1695. The entablature of decorative fifteenth-century flint flushwork was reused around the wall base; to the right of the window appears the signature of master mason, Richard Russell, who was contracted to build the tower in 1426.

in England. However, the churchwardens' accounts date from slightly later in the mid-fifteenth century. In 1454 there was payment for carriage of stone from the old church, suggesting demolition. This probably continued piece-meal for there is a further entry in 1474: 'paid to John Lewke for taking down of the windows.' Both Gardner and Lewis interpreted these entries in different ways, but the Agas map and the seventeenth-century rentals which refer to the 'Old Church' indicate that there were still some visible remains long after the new church was completed.

The churchwardens' accounts give us a vision of the workings of the pre-Reformation church and community. However, Gardner's interpretation of the accounts needs to be approached with caution. For example, under 1452 he mentions St Andrew's Foot, which is in fact St Andrew's 'fest' or feast. He quotes an entry from 1453, a year which is in fact missing from the accounts; his entry reads 'payd for nettys for the lowe awters.' This must be his attempt to transcribe the entry for 1456, which follows on from the entry for 1451 '… for ii mattys for the lowe autyre.' By the standard of his time, Gardner was a good antiquarian, but his palaeographic skills were sometimes wanting.

It is very clear from the published receipts that apart from the substantial income that derived from 'fishing fare', itself difficult to assess since it was frequently paid in a mixture of fish and cash, ordinary parishioners strove to contribute towards their new church in the time-honoured way. Three festive church 'ales' in 1453 raised in total 32s. 8d. Money gathered in the church raised another 7s. 2d. In 1458 four separate Sunday gatherings raised an average of nearly sixteen pence each. Wives who collected in the town in 1496 raised 10s. The Guild of St John was a useful contributor; on the Sunday before his feast and on the day itself 4s. 4d. was given. There were many small sums of money given for the salvation of souls, 'Romescott' and pascal gifts were also regular sources of income, but occasionally larger sums were paid for the salvation of the souls of wealthy individuals. For example, 6s. 7d. was given for the soul of Sr Harry Barbour and 13s. 4d. for the soul of Agnes Schobote in 1458. The shop at the church gate also brought in a few pence. The impression given is very similar to that found in other late medieval parish churches, of a lively community willingly contributing through gifts and offerings and the regular payment of 'fishing fare' towards a very materialistic church.

The church is believed to have been consecrated in 1492, when there is mention of money paid for the bishop's dinner, however, lime and floor tile was still being purchased and the bishop was entertained again in the following year and again in 1497. Episcopal visitations, in person or by representation, were normal annual events and may or may not have coincided with a dedication ceremony. It is in fact uncertain when the church was finished, payment to a mason in 1499 for work on the stairs to reach the candle-beam suggests that the finishing touches continued up to the turn of the century. Walberswick church, although enjoying all the sacraments, remained a

dependent chapelry of Blythburgh and its tithes, including the valuable tithes of fish, were the property of the Augustinian Priory of Blythburgh. When the priory was dissolved in February 1537 these tithes fell into secular hands and eventually became part of the lordship of the manor. With the Reformation, all the tithes, small gifts and offerings slowly dried up and the churchwardens found themselves become more and more involved with secular matters, appointing overseers of the poor, collecting rates and doing their best to make ends meet. These changes probably coincide with a significant shift in the religious opinions of the parishioners away from the medieval world of iconography and superstition towards that of the new ideology of the Reformation and Protestantism.

There is no reliable way to assess the religious opinions of individuals in the past, least of all a whole town population, but there are indications as we get into the early seventeenth century that Walberswick was becoming an increasingly puritan community. Spufford talks about the 'iron curtain' that obscures religious opinion among ordinary parishioners, but she has also used the religious professions in the preambles of wills as a useful indicator of religious belief. Any mention of intercession, or the Blessed Virgin Mary, or the Company of Heaven, is likely to indicate Catholic sympathies, whereas an emphasis on personal salvation and Christ's Passion might suggest Puritan tendencies. Individuals rarely wrote their own wills, but they were unlikely to have a will written for them by someone whom they did not trust. These final testimonies can serve as a useful rule of thumb for religious affiliation.

There are a small number of published wills from Walberswick for the early seventeenth century and they give some insight into personal belief in the community at that time. For example, Edmund Gardiner, yeoman farmer of Walberswick in 1620 commended his soul to: 'Almighty God maker and redeemer'. More humble but assertive was William Copping: '... believing I will be saved by the blood shed and obedience of Jesus Christ.' His body was buried inside the church in 1621. Nicholas Casson of Walberswick, gentleman, made his will in 1621 asking to be buried at Walberswick believing he would be: '... saved by the merits and passion of Jesus Christ redeemer.' His daughter Catherine had married into the Cornish Tradescant family who held lands in Blythburgh, Wenhaston and Thorington in 1635 as well as in Walberswick.

More individually puritan in style was Christian Smith, a widow of Southwold, who left a messuage, lands and tenements in Walberswick and 20s to the minister at Southwold: 'seeing that life is like a flower of the field that soon withers ...' Shipwright and Walberswick resident, John Chettleburgh, gave his, 'soul to the hands of Almighty God, creator,' in 1625. He was the brother-in-law of Henry Richardson, to whom Sir Robert Brooke leased the quay in 1630. Sebastian Mowling, who lived in the Town Tenement at about the same date, asked to be laid in Walberswick churchyard and commended his soul to God, through Christ by whom he hoped to obtain remission of sins and to look for everlasting life. All these examples tend towards the puritan end of the religious spectrum and some are overtly puritan.

One of the most interesting published wills is that of Thomas Barweck, fishmonger of Ipswich, who made his will in 1623. He was the brother of John Barweck, almost certainly the same person as the John Barwick, churchwarden, book-keeper, the man who recorded the *Bloody Marsh* incident for posterity, and chief protagonist in the struggle against John Brooke. The preamble to Thomas' will is exceptionally long and strongly puritan in character: 'Being an unprofitable servant of God ...' and 'making him a reasonable and living creature ...' etc. he held a half share with his father,

Ewen, in a salthouse at Walberswick which he left to his only daughter Mary who was under 18 years of age in 1623. The family of his brother John would inherit on her death. There was another brother, Roger, and two sisters, Mary and Margaret. To Ewen his father he left his 'sorrel ambling mare'. This was an important family in Walberswick, unfortunately a complete family reconstitution is impossible because the parish registers do not begin until 1656, however the will of Thomas and the writing of his brother John suggests that this was a strongly puritan family. John was elected churchwarden on a number of occasions from 1627 onwards.

Of the dozen or so published wills that mention land in Walberswick, all were either puritan or even overtly puritan in character, but it is unlikely that all the inhabitants were of the same opinion. If there were any Laudian or Catholic sympathisers in seventeenth-century Walberswick, they very wisely kept a low profile. However, there is one person who was declared a 'delinquent' and whose property was sequestrated. A certain 'Henry Fearmes' was made churchwarden in 1614. In subsequent years there were problems over payments made on the doles due from his fishing boats. Over several pages his entries in the accounts have been corrected in a different hand. The implication is that he was fiddling the books.

In 1632, there was an order for the sequestration of a 'Henry Ferns' property. In 1637 his estates, or whatever remaining title he may have had in them, were surrendered to his mother and brother Joseph and his heirs. His sisters seems to have inherited them and they in turn sold out to John Barwick (junior) and Elizabeth Chapman who were acting as trustees for William, the son and heir of John Chapman – another protagonist against Sir Robert Brooke in the 1640s. In 1653 the claim to these estates by William Chapman's trustees was allowed. Another Henry, probably the son of Henry Fearms was the fellow churchwarden of Henry Richardson in 1644, and was said to be responsible for removing one of the parish boundary markers (Document 16).

The story of Henry Fearmes highlights not only the intricacy of family relationships in seventeenth-century Walberswick, but also the importance of material considerations. Henry was probably not declared a 'delinquent' so much for his religious beliefs as for his abuse of office as churchwarden. Sequestration was the most effective punishment available at the time for a churchwarden who was prepared to lie about the money he should have been paying for his fishing dole. Considering the impoverished state of the town in the 1620s, we can only imagine the outrage of the community when a dishonest churchwarden was discovered in their midst.

The Civil War made its impact on Walberswick, as it did in nearly all parishes, both intellectually and physically. In 1644, the church was visited by William Dowsing and his troopers; small sums were paid for the taking down of 'painted images' – including no doubt the stained glass windows, removing brasses with sacrilegious inscriptions which were then sold for scrap, and for taking down the rails before the altar. In accordance with the orders of Parliament, the King's Arms were erased from inside the church in 1650.

Laconic entries in the churchwardens' accounts give few clues as to how such changes were viewed by ordinary parishioners. In 1586 the Archbishop of Canterbury's visitors requested the churchwardens to repair a copy of the 'Paraphrases of Erasmus' and to provide Masculus's 'Common Places' or the 'Apologies', such books were required reading and had to be kept in the churches of the diocese. The Erasmus was sent to London for rebinding at a cost of 3*s.*, so no doubt it was well used.

Copies of the 1551 edition of Erasmus' 'Paraphrase of the Gospel' survive at Bramfield, Chediston and Sotterley parochial church libraries, while Jewel's 'Defence of his Apology', published in 1567 was also recommended reading in 1610. A brief inventory of books in Walberswick church, dated 1609–10, included a large bible and a prayer book, other 'black books' including the Erasmus. So the great Renaissance humanist and satirist of Catholic religious observances was not just compulsory reading, but probably an old favourite as well; perhaps we underestimate the degree to which ordinary parishioners were informed about the theology of their day. In 1653 there is mention of a 'meeting house' at Walberswick, but this was well before the licensing of Independent meeting houses. John Barwick also mentions 'our church or meeting place' in 1652 as if they were one and the same building (Document 13). Certainly the churchwardens were spending money on a 'meeting house', but it is not clear therefore whether it represents a break-away non-conformist group, as Gardner implied.

At Southwold, both the Independents and the Anglicans used the same parish church together, services followed on one from another, with many people staying for both. However, the puritan enthusiasm of the inhabitants at Walberswick is not in doubt, for in 1654, when Nathaniel Flowerdew was minister, it was reported that they: 'are much taken notice of their zealous affection to the Gospel, and to the cause of the Commonwealth, having had sundry able ministers among them, and supported them without any assistance from the state'. The Flowerdews had been prosperous puritan mercers at Walberswick in the early part of the century. In the years immediately before 1654, the minister Mr Stephen Fen, a 'faithful Servant of Christ', served both Walberswick and Southwold parishes. Both Fen and Flowerdew are missing from the list of ministers given to us by Gardner, which is substantially incomplete, even though Gardner tells us about them. It is possible that at certain times there was both an official vicar appointed by the lord of the manor and a separate puritan 'minister' who was chosen by the community.

After the reformation, the vicars were appointed by the lord of the manor and the living was shared with Blythburgh. This is probably why the vicars never took sides in the protracted disputes with the Brookes over common rights. Indeed, these poor men must have walked a political tight-rope between loyalty towards their embattled and impoverished parishioners on the one hand, and the litigious and belligerent squires who employed them on the other. The tithes were leased out to local farmers and the ministers were given a fixed stipend; this was a bone of contention in the mid-sixteenth century.

FIGURE 32. The lych gate and ruins of St Andrew's church (chapel) Walberswick. In June 1695, faced with a declining population and grinding poverty, the church wardens ordered the 'lessening' of the church, which had once been the pride of their town. Twenty tons of lead from the roof were shipped to London; with the proceeds a small chapel able to hold just '40 worshippers' was constructed using materials from the old church.

John Barwick complained: 'whereas he [John Brooke] saith the minister hath the tythes, that is not soe, for he hath let them to one of his cheif fearmers within our town, who have had them formerly, and very stricktly he then gathered them, and now will do no lesse, but have what he pleaseth' (Document 14).

The farmer was then supposed to pay the vicar £20 per annum for his stipend, but it was not always forthcoming. In consequence some vicars left the living and there were periods when Walberswick had no vicar at all. After the death of Squire John, 'the great troubler', Elizabeth Brooke is said to have given most of the tithes from the chancels of Blythburgh and Walberswick churches 'to him that had the care of souls', reserving little for building repairs. At least some of the vicars in the second half of the seventeenth century may have found a friend; let us hope that there were occasional visits to Cockfield Hall for tea with Elizabeth, where she might indulge her passion for theological debate – an intellectual oasis perhaps – if not too daunting for younger clergymen.

Elizabeth's failure to maintain the vast crumbling medieval structure of Walberswick chapel was another crisis waiting to happen in the later part of the century. In June 1695, the churchwardens accepted the inevitable and an order was issued for 'lessening' the church. The three remaining bells were sold – two were cracked anyway – the lead roofs of both chancel and nave were stripped, the timbers removed and the whole of the central aisle and nave clerestory was demolished. Twenty tons of lead were shipped to London and fetched over £160; two of the bells fetched another £50 and other bits of timber, iron and lead were sold locally raising altogether £303 1s. 11d. With some of the salvaged building materials a small chapel, large enough to take about forty worshippers, was constructed inside the south aisle costing £290 8s. 9d. The churchwardens had £11 13s. 2d. in hand as a result of this contraction; this forms the final entry on their accounts for 1696/97. Some memory of Richard Russell's work in the 1420s must have lingered on, because the massive parapet stone bearing his mason's mark was carefully placed beside the new east window: a reminder perhaps of a time when temporal and spiritual wealth were combined successfully in a mighty edifice.

God and Mammon: chapter notes

George Crabbe, 'The Dumb Orators', lines 269–78. Although it relates to Justice Bolt, 'impetuous, warm, and loud', this passage seems to fit the shy vicars of Walberswick, sitting on the rickety fence of their livings. The Blythbugh Priory Cartulary: Harper-Bill 1980; tithes of the sea in 1411: *ibid*. pp. 238–40. Flooding in the fourteenth century: Baker 1966; Mortimer 1979, p. 7; Warner 1982, p. 16. Medieval pottery found on the beach: *PSIA*: 29 (ii) 1962 p. 173; 31 (i) 1967 p. 82; 31 (iii) 1969 p. 329. Quotations for building work: Lewis 1947, p. iv, but see also Gardner, p. 152. Money raised for the church: Lewis 1947, pp. 88, 99, 256. Bettey 1979 captures the material character of the fifteenth-century parish church in England. Religious affiliation attested in wills:

God and Mammon Spufford 1974, pp. 320–7. The wills quoted in this chapter are published in: Allen, 1989 and 1995. Gardner, p. 163, lists persons buried inside the church. Tradescants of Walberswick, famous gardening family: Allan 1964. See also Clare 1903 for Tradescants in Wenhaston and Thorington. Lease of the quay to Henry Richardson see: SRO, HA30: 50/22/27/3.2. Thomas Barweck's will: Allen 1995, p. 485. John Barwick elected church warden in 1627: CWA, fol. 131. Henry Fearmes elected church warden in 1614: CWA, fol. 103. Henry Fearmes sequestration: State Papers, Calendar for Compounding etc. 1643–1660, Part I–V, 2820. For the visit of William Dowsing see: White 1883/88; Gardner p. 160. Rebinding of the 'Paraphrases of Erasmus' in London: CWA, fol. 18r. Local parochial church libraries: Fitch, 1967. Brief inventory of books at Walberswick in 1609–10 see: CWA, fol. 99. Gardner mentions the Southwold Independents at some length, p. 212. The report of 1654: Gardner, p. 167.

CHAPTER SEVEN

Accusations of rape

....

Now here was Justice Bolt compell'd to sit,
To hear the deist's scorn, the rebel's wit;
The fact mis-stated, the envenom'd lie,
And staring, spell-bound, made not one reply.
Then were our laws abused – and with the laws,
All who prepare, defend, or judge a cause:
'We have no lawyer whom a man can trust,'
Proceeded Hammond – 'if the laws were just;
'But they are evil; 'tis the savage state
'Is only good, and ours sophisticate!'

Crabbe

William Turrould was a brave man to take the lord of the manor to court over common rights, but he himself may not have been without fault. Not everyone was on his side; men like Henry Richardson, Richard Girling, Thomas Palmer, Francis Colbie and the Whincopp brothers, Lawrence and Edmund, who had taken up the lease of Westwood Lodge farm (excluding house and park) from Sir Robert Brooke in 1614, had a vested interest in the new enclosures. The Whincopps took over the lands of Westwood Lodge 'unto Walberswick tunmere' with 'all those inclosed marshes with a wall or bank', but they had to maintain the marsh walls. The Whincopps were substantial yeomen farmers, paying 20s. more in Ship-money to Blythburgh than Sir Robert Brooke himself. Lawrence came from Weston while his brother Edmund had lands in Middleton and Theberton.

Palmer and the Whincopp brothers were men apart, distinct from the impoverished Walberswick community. The churchwardens took them to court for refusing to pay poor rates for their land in Walberswick; in January 1638 the justices of the peace at Beccles ordered them to pay up. Turrould was the Sexton of the church at Walberswick and so we can begin to see divisions within the community, between the older established customary tenants led by the churchwardens and parish officers, and Sir Robert's new farmers and their employees and labourers.

On 19 December 1636, Robert Tanner of Walberswick made a lengthy statement about an incident that had happened more than a year earlier. Apparently, towards the end of November 1635, William Turrould had arrived at the house of Robert Edmunds in Walberswick in the evening, went inside and shut the door behind him. In the house at the time were Margaret, wife of Thomas Sherring, and Alice, the wife of William Sallows and daughter of the owner of the house, Robert Edmunds. The only other occupant was Alice's baby. The Sallows family may have been one of the Brooke's farmers since they gave their name to Sallows Walk, a wide strip of heathland adjoining Westwood Park. Tanner does not give any information about what was said, or indeed, if there was any conversation at all. His silence speaks volumes (Document 3).

Turrould then attacked Margaret Sherring: 'then and there did with his yard drawne violintly put his hand under the clothes of the said Margaret Sherringe and did inforce hir to have comitted ffilthines with hir: Who striving with him to hir power yett could not ffre hirself from him called to the other woman for help: She laying hir infant which she had in her armes upon the harth did with sped runne to help the said Margaret Sherringe.' In the ensuing struggle Turrould then turned his attentions upon Alice, getting her down on the ground where he 'did strive also with hir in so much that she was constrained to call out for help.' The mother of Alice Sallows, who was outside, hearing the commotion, came rushing in. Failing to pull Turrould off her daughter, she beat him with a staff, 'therewith before she could make him forbeare this abominable acte' (Document 3).

The story has a sense of melodrama, and in the intervening year, no doubt, it had lost nothing in the telling. What is missing from the statement is most interesting. Was this an unprovoked attack? Was Turrould really seeking sexual gratification, or was this a 'domestic' row between neighbours that had got out of hand? Without doubt there were underlying issues behind the accusation, in particular Turrould's role in leading the townsfolk in their campaign for grazing rights, but he may also have been an unpleasant man. None of the persons mentioned in the statement appear on the 1636 rental for Blythburgh and Walberswick, that is to say they were not copyhold or freehold tenants of the manor. Neither do they appear on the Ship-money returns for Blythburgh in 1639–40. One should not read too much into negative evidence, but the implication is that they were probably landless labouring families working for the new leasehold farmers, and therefore beholden to Sir Robert Brooke.

Turrould was brought up before John Scrivener, Justice of the Peace, who, based on the statements of the three women, which we are told, 'was rather more vile than is above spasefied', bound him over until the next quarter sessions. But then the case was thrown out by Henry Coke JP of Thorington, a neighbour and arch-rival of Sir Robert Brooke, who declared that the women: 'ware none but roagues and whores and theves that cam against him'. In consequence no action was taken against Turrould. Of course, nothing in this story can be taken at face value. It has to be seen not just against the background of petty feuding among the peasant families of Walberswick, but against the rivalries of neighbouring landowners.

The partisan nature of the two JPs places the incident in the wider world of county politics. John Scrivener's hand and signature survives in the Brooke archives on a statement which he took down from witnesses who were trying to discredit Turrould. It is possible therefore that he was a supporter of Robert Brooke. Coke, on the other hand, was undoubtedly an enemy of the Brookes. Not only had Henry Coke challenged Sir Robert's right to lease out the quay at Walberswick, but he had also led the opposition in the ensuing Exchequer court proceedings. Sir Edward Coke had purchased a 99-year lease of the manor of Hinton from the Hoptons. The manor had originally belonged to Blythburgh Priory and the lease had been drawn up just before the Dissolution. Subsequently all the priory lands had been granted to Arthur Hopton, but the lease still had some time to run. Meanwhile the Brookes purchased the Hopton estates. When the lease ran out, Henry Coke remained in possession and assumed the title of lord of the manor, retaining the court rolls and issuing copyhold tenancies, which the Brookes believed to be invalid. Sir Robert Brooke fought for many years to recover his possessions in Hinton and he eventually succeeded in getting compensation. There were another thirty-seven acres in Westleton from which Henry Coke received rent, but which Brooke also believed to be his. This argument was raging at the level of the king's council in 1638, but was not resolved until twenty years later when Coke gave Sir Robert a bond for an award of £180. The Cokes and the Brookes were therefore at daggers drawn; William Turrould was simply the happy beneficiary of that fact.

Turrould presented his first petition at Greenwich on 23 January 1637. It must have taken a while to prepare, so the timing of the accusation of rape in December 1636 is critical. It also explains why the accusation was made nearly a full year after the alleged event. Was this really a rape at all, or just some trumped-up charge, an elaboration of an argument between neighbours? There is no doubt that Turrould was a forceful character and a man to be reckoned with, but there were other accusations against him which suggest that there was a darker side to his character. If Sir Robert was looking for material to discredit Turrould, he did not have to look very far.

Three more statements fit into this same category of mud-slinging against Turrould, but if anything, they are more substantial than the accusation of rape. Furthermore they are clearly related to the arguments about enclosure. There are some minor differences in detail, but two accounts tell essentially the same story and a third, which is unsigned, may be little more than malicious gossip. The statements made by Thomas Mayhew and George Cricker, a minor landholder in Blythburgh, are first-hand and clear enough: on 31 October 1636 they were working on a bank near the mill of Walberswick when they were spotted by William Turrould, the miller, who was walking from his house in Walberswick towards the mill. Going into the mill he fetched out a quarter-staff, 'having graynes in the one end and a pike in the other and a sword by his side.' He approached the two workmen and said: 'I forfend you working any longer heere, if you doe I will kill you if there were no more men in England.' Then he made a thrust at Thomas Mayhew with the staff and delivered, 'five or six blowes with the said staffe, with which blowes he broake his said staffe and when he had done soe he drew his sword at the said George Cricker and swore by the light of God that if we or any other man, aye if it were our great Master Sir Robert Brook came there eyther by night or day he would shoote a bullett in his sides, if he were hanged within an hour after' (Document 7).

If it is to be believed, this was dangerous language indeed, and the accusation a serious one. The idea of tenants attacking their squires would have been seen as insurrection in the early seventeenth century. Kett's Rebellion in Norfolk, which had taken place less than one hundred years before, was prompted by enclosures and had involved attacks by the peasantry upon their squirearchy. It was savagely supressed and several hundred of the rebels had been executed. There had also been similar violent disturbances in Northamptonshire over enclosures in 1607, which were seen as a serious threat to public order.

Another version of these events was given by Robert Tanner with some embellishments on Turrould's character, it has all the flavour of local gossip and tittle-tattle. We are reminded that William Turrould had lately abused Margaret Sherring: 'and did endeavour to have ravished her using great violence upon her and threw her down to the ground and would have – (done his pleasure upon [crossed out]) – abused her had he not byn prevented by company that was called in'. That Turrould had walked up and down the town of Walberswick with his sword by his side: 'and will not suffer any execution of [presse] to be done upon him, to the great terrors of the poore inhabitants of Walberswick.' Turrould, we are told, was so much subject to drinking, that in his drink he used 'vile language' while spoiling for a fight and intimidating his neighbours. After the attack on Thomas Mayhew, his assailant: 'bragged and bosted that he had payed this defendant his wages and would pay him again soe that he should bee ware' (Document 7).

Historians have long been aware that the concept of public order is inextricably intertwined with local issues, squabbles and vendettas; the local constables decided whether to prosecute or ignore offenders taking into consideration their own personal safety, friendships, family and political loyalties. Descriptions of misdemeanours such as these are typical of the seventeenth-century village community. They are likely to be highly coloured, and they leave out elements which would be essential if one was to come to any rational understanding of what had happened, or why the case had been brought in the first place. We are not told why the two men were working on a 'bank' or what its purpose was, yet given the context we can be sure they were busy on Sir Robert's enclosures. Typically we only hear one half of the conversation. Turrould cannot have been the first man to have got drunk in Walberswick, or to swear in his cups; we only hear about this incident because Turrould was pursuing an action against Sir Robert Brooke and there were members of the local community who were not happy about it.

Turrould may have escaped lightly when Henry Coke threw out the case against him in 1637. However, his enemies got the better of him two years later, for an entry in the churchwardens' accounts stated that: 'William Tourrold our sextonne being now prysner in Blythborough jayle our church-wardens ... at our church in Walberswick this 14 daye of December 1639 ... then placed the said Francies Coale sextonne and delivered him the keys and other things till Easter day next ...' In the following year there is an entry in the Quarter Sessions Order Book concerning the 'House of Correction' and the jail at Blythburgh, 'being in great decay'; £15 was provided to the chief constable of the Hundred for its reparation. At the same time, Humphry Hearne: 'whoe is now in possession of the said house of correction at Blythburgh', and who had presumably been Turrould's jailer, was relieved of his post. Furthermore: 'William Tirrold of Walberswick miller is moated and appointed by this court to be keeper of the house of correction at Blythburgh for this division ...' Hearne was then ordered to deliver up the keys to Turrould, although he could reap the crops that he had already sown on the ground attached to the jail.

This statement provides an interesting insight into Turrould's character, because the job of looking after the hundredal house of correction in any seventeenth-century community was not one for persons of a lovable dispo-sition. Debtors and petty criminals, murderers and lunatics, alcoholics, the dispossessed, the halt and the lame, all were collected into these hell-holes of social deprivation. The 'guardians' or 'overseers' were notorious bullies, cruel and brutal, often using and abusing their charges as slave labour to supplement the meagre payment they received for their office. Indeed, in many cases it was assumed that the overseer would lease the jail for that purpose and might even pay for ready access to cheap labour. Let us hope that, for the sake of his soul, William Turrould was one of the kinder sort who took their responsibilities seriously, but given what we know about him, this is unlikely. It is possible that he sub-let the land attached to the jail for a 'Umpherye Thurston' was paying ship money for the the 'Gaole grounds' in March 1640. Two years later in 1642–43, the keeper of the jail and house of correction at Blythburgh was once again Humphry Hearne; it was then William Turrould's turn to hand back the keys. Such an extraordinary reversal of fortune seem worthy of a Thomas Hardy novel – but these were extraordinary times.

The fact remains that in 1637 Turrould led the majority of the inhabitants of Walberswick in their legal challenge to stop Brooke enclosing the commons. What sort of man was Turrould? We know he was the miller of Walberswick and the mill belonged to the town and not the lord of the manor. Sir Arthur Hopton had provided an acre of land for a token ground rent and given it to the town in 1529, so they could build a mill and mill cote there and receive the income from it. From other documents it is clear that the mill then stood about a quarter of a mile out of the settlement on the modern Blythburgh road near what is now Eastwood Lodge farm (Map 1). A windmill of the nineteenth and early twentieth century stood in the Mill Field, now within the modern village of Walberswick.

The miller probably had some social standing in the community and

Turrould was sufficiently respectable to become church sexton and involved in local affairs. He is mentioned as church sexton in a note about the beating of the bounds of Walberswick in 1644. So both his time in jail and being jailer had not diminished him in the eyes of the inhabitants, who must have known more than we do about him and his circumstances. At Deadman's Cross there once stood an ash tree: 'whereon were made hundreds of marks by the townsmen of Walberswick in old times when they went their bounds.' Henry Richardson: 'one of the antientest and chiefest men in Walberswick, reported at Deadman's Cross, in the presence of more than forty souls, that he was one of the church-wardens when that tree was fetched ... and laid into the church of Walberswick.' Fourteen years later the churchwardens ordered the stump to be split up for firewood, 'fireing scarce', to melt lead for the plumber working on the roof. In 1644, 'William Thorrold', the church sexton, confessed to having destroyed this important relic (Document 16).

Such details seem inconsequential now, but they were sufficiently important to be recorded by the townsfolk at the time, for they determined the ancient boundaries through the upland commons between Blythburgh and Walberswick, which was central to their argument over grazing rights. Turrould cannot have been ignorant of the importance of this tree and why it was in the church for safe-keeping. As sexton it was his responsibility. This story gives us yet another vision of William Turrould, as a short-sighted, not to say dim-witted member of the church hierarchy. This vision is not difficult to reconcile with his rash and extreme behaviour in the articles of misdemeanour and the accusations of rape in 1636, his time in jail three years later, and then prisoner turned jailer in 1640, accepting a job that most people would have avoided. William Turrould was certainly no under-dog. He was a tough, bold and outspoken man, a bully perhaps, cruel and hard, the sort of person who found it difficult to keep out of trouble – in fact, the perfect rebel.

Accusations of rape: chapter notes

George Crabbe, 'The Dumb Orators', lines 191–200. Ship Money returns for Blythburgh: Redstone 1904, 75 (Walberswick was exempt as a coastal town). Whincopp leases of 1614 and 1621: HA30: 50/22/1.12. The nineteenth-century tombs of the Whincopp family can be seen in the church-yard at Blythburgh. 'Sallows' could also indicate willows, but the dry heathland walk here favours interpretation as a personal name; there was a John Sallows, thatcher, at Walberswick in 1794: HA30: 50/22/3.37. There are many documents relating to the dispute with Henry Coke among the Cockfield Hall papers, but two in particular are relevant to this chapter: SRO, HA30: 50/22/3.1 [16]; HA30: 50/22/3.21 [11]. There are three similar statements concerning the misdemeanours of William Turrould among the Cockfield Hall papers: SRO, HA30: 50/22/3.1 [47, 48 and 52]. Quotations have been taken from all three, but only [52] has been selected for publication as Document 7. For different concepts of public order in seventeenth-century England see: Wrightson 1980; Underdown 1985). The statement about William Turrould being in jail in 1639 comes from the church wardens' accounts: CWA, fol. 143, while the statement about him being made jailer comes from the

Quarter Sessions Order Books: SRO, B,105/2/1,16, and Redstone, 1904, p. 75 for the Ship Money Returns. For the return of the jail keys to Humphry Hearne see: SRO, B. 105/2/1, 53. The clearest indication of the location of Walberswick windmill comes in statements concerning the location of Black Hill in William Howlett's case: SRO: HA30: 50/22/3.1 [57] Here Black Hill was said to be, 'a little beyond Walberswick wynd mill, on the north side of the said town' and that 'Black Hill ... is above a quarter of a myle from Walberswick townes end on the north syd thereof.' The ground falls away steeply a quarter of a mile north of the town and would be unsuitable for a windmill, but to the north west there is high ground and Eastwood Lodge is exactly a quarter of a mile from the last house on the 1582/83 rental.

CHAPTER EIGHT

Rogues and neighbours

> He was a member of a proud and powerful aristocracy,
> and was distinguished by many both of the good and
> the bad qualities which belong to aristocrats. His family
> pride was beyond that of a Talbot or a Howard. He
> knew the genealogies and coats of arms of all his neigh-
> bours, and could tell which of them had assumed
> supporters without any right, and which of them were
> so unfortunate as to be great grandsons of aldermen.
>
> *Macaulay*

It would seem that Sir Robert Brooke and his son John bear all the characteristics
of 'rogue' squires; indeed John Brooke has been described by one modern
writer as the 'wickedest of country squires'. Like Colonel John 'Blackbeard'
Mohune, who haunted the churchyard in search of treasure in Meade Falkner's
'Moonfleet', John Brooke, the 'great troubler' is said to haunt Westwood Lodge.
His ghost known as 'Old Fleury' rides madly down the passages, they say,
sitting on a saddle; in 1865 a saddle, boots, riding whip and spurs were discovered
concealed in the house. 'Rogue' squires abound in eighteenth and nineteenth-
century literature; although under-studied, they are one of the formative themes
in the history of the British novel. Henry Fielding's innocent Joseph Andrews
encounters numerous abominable but nameless squires – nameless for fear that
Fielding's readers would recognise themselves. There were of course some
notable genuine 'rogue' squires who served as models for the genre: John
Murgatroyd and his sons of East Riddlesden Hall, Keighley, in Yorkshire, for
example. The river Aire is said to have changed its course in the 1660s at their
doings; there was then 'a rot among the gentry' according to a local minister.
Gough's history of Myddle, Shropshire, in the early seventeenth century was
not short of rogues either, at all levels of society. Closer to home, Richard Kirby,
the dreadful self-styled squire of Landbeach, who nearly sparked another Kett's
Rebellion by overstocking the fenland commons and impounding tenants'
cattle, was, like the Brookes, one of the new gentry. Later classic novels such as
'Moonfleet' and 'Laura Doone', played on such stories and offered irresistible
stereotypes to which nineteenth-century local historians were drawn. When
considering the Brookes and their tenants we must take care to assess them fairly
by the standards of the day and not be influenced by later romantic literature.

At first sight the case against them seems strong. Many of the antiquarian sources following Gardner make reference to the malicious actions of the Brookes, particularly John Brooke. This is also backed up by some of the primary sources, such as the petition of William Turrould, which refers to 'cruel and unjust dealings', but Turrould was out to make his point. The town had 'groaned', so we are told, under the burden of Sir Robert's lordship. John Barwick, had good reason to vilify his landlord having been arrested a number of times by John Brooke's bailiffs. Barwick was particularly venomous in his complaints: 'a very great troubler and an oppressor of his poore tenauntes in Walberswick.' Brooke was 'envious'; his actions were 'wicked' ... 'troubling and sewinge of poore men' ... 'exstreemely oppressinge and tearyfyinge of his pure tenaunts' (Document 13).

Gardner was a good eighteenth-century antiquary, but his approach was hedged around by a concern not to offend any of his subscribers; the list of whom at the front of his book includes two members of the Blois family, descendants by marriage of the Brookes and then owners of Westwood Lodge and Cockfield Hall. Other subscribers included Sir John Rous of Henham, the Barne and Bence families, important local landowners, all of whom had paid in advance for copies of Gardner's work. Gardner had to be careful not to say anything which might reflect adversely on the descendants of his historical subjects. Subsequent writers found it easier to use the printed word of Gardner rather than investigate the primary sources, most of which were hidden away in remote corners of Cockfield Hall.

Of course, the archetypal 'rogue' squire has his counterpart in the benevolent 'good' squire, epitomised in Fielding's 'Squire Allworthy' in Tom Jones. Barwick contrasted the time of Brooke's lordship with that of 'Sir Owen Hopton's days'. Hopton was 'a worthie jentleman and loved the poore town': he was clearly also implying that the Brookes were not gentlemen. We are even told Sir Owen went down to Pauls Fen and asked the town neathers's boy to come up to the lodge for a drink (Document 14). So too was the 'virtuous Lady Brooke', Jane Barnardiston, widow of John Brooke, who ran the estate between John's death in 1652 and her marriage to William Blois in 1660; she allowed her tenants to take up their common rights again (Document 14).

The image of the lovable 'Squire Allworthy' in an idyllic 'Paradise Hall', is an older and even more powerful literary stereotype than the 'rogue' squire. Ironically the Brookes were distantly related to the Fastolph family, whose name became equated with the archetypal 'good' squire. Images of a Shakespearean Falstaff are echoed in Robert Reyce's 'Breviary of Suffolk', where the gentry: 'converse most familiarly together' winning one another's goodwill. Here is yet another dangerous literary formula which has influenced modern historical writing; Reyce has become accepted as a historical source in its own right, yet in some ways is as exaggerated and biased as the vitriol of Brooke's tenants. Such powerful literary stereotypes must be banished temporarily from the reader's mind while we study the real characters from seventeenth-century

Walberswick; the reader will not be disappointed, for some are larger than life and would fit comfortably into one of Fielding's novels. So, were the Brookes truly 'rogue' squires?

There is a mystical twist in the historical writing of Janet Becker which deserves digression. She was the daughter of Harry Becker the artist and grew up at Blythburgh clearly loving the place. Although not an academic historian, she used her knowledge of the locality and whatever sources she could find to good effect. Clearly she was puzzled by constant reference in documents to the Brookes as villains of the piece. She experienced a dream before starting to write her book and this vision she included in the appendix. In her dream she met and conversed with Squire John Brooke. She describes him as: 'a tall, wiry young man, dressed in black and red clothes of the Cromwellian period'. She questioned him about his enclosure of the commons and the fact that nobody spoke good of him. In her dream the answer came back:

'Has it ever struck you that there may be two sides to a question?' (note the interesting lack of seventeenth-century syntax!).

'Do you know that the Hoptons sold us so many acres including the commons? They sold us what was not theirs, and when my family took over, the people began filching here and pilfering there till we would have had nothing left of what was ours by right of purchase if we had not fought for it.'

FIGURE 33. The south front of Westwood Lodge recently reconstructed by Mr E. T Webster. The house is famous for being haunted by the ghost of the 'great troubler' John Brooke; 'Old Flurey', it is said, charges down the corridors on a saddle.

'I see', said Janet, 'this certainly is a new side to it.'

Becker was well aware that dreams are not the concern of serious historians, yet many a dream has changed the course of history, and, it would seem, historical writing also. In fact Becker's dream was an ingenious device, through it she expressed her awareness that the sources were biased against the Brookes, but she could not prove it. She knew that arguments over the commons dated back into the Hopton era, but she had no other sources to guide her. Sixty years on, there are many more sources readily available and more is understood about seventeeth-century economy and society. Now we are better equipped to understand the Brookes from the inside, their social and political background, the decisions that they made, their economic needs, and the prejudices which they were likely to encounter. Thus we may be able to redress the balance, or at least understand their situation, even if we do not sympathise with them. Above all we must assess their circumstances in relation to their social and economic standing in the seventeenth-century county community. If we enquire of their family and friends, then perhaps we will be better equipped to understand their critics and their enemies.

Alderman Robert Brooke purchased Blythburgh, Walberswick and Westleton manors from Arthur Hopton in 1592 for the sum of £5,200. Janet Becker has told the story of the Brooke family, which need only be summarised here. Almost immediately after acquisition, litigation commenced, with Hopton accusing Brooke of fraud by 'interlining' the names of other sub-manors in the purchase agreement. This was a bad start, for the Hoptons had been popular. Thereafter the Brookes seemed to attract litigation which spanned three generations of their lordship at Blythburgh. It is not easy to explain why. Were they perceived as an intrusive gentry family by their neighbours? Or was their litigiousness part of an urban elite culture that they brought with them from the city? Only they could answer such questions. Alderman Robert Brooke was a successful London grocer and had all the appearances of a self-made man seeking gentry status in the countryside. The stigma of a background in 'trade' was not a barrier to gentility then as it was during the period of the pseudo-gentry in the eighteenth and nineteenth century; it was not 'unfortunate' to have an alderman in one's ancestry in the seventeenth century: Macaulay's whiggish comment merely expresses the prejudice of his own day.

This was the period of a rising gentry: men who had made money through trade in the city or in the new professions and who sought the time honoured status of a country seat and a coat of arms. Becker, drawing on Copinger, says that, Alderman Brooke's father, Edward Brooke, was already a country gentleman from Aspall, so if he had family roots in Suffolk, it might explain why he sought property in the county. Gardner was probably right in making a distant connection with Lord Cobham. Nevertheless, from all appearances, he was 'trade', but even in that respect he was no different from many of his neighbours who had fingers in various pies, in commerce, government and politics. The Rouses of Henham, who bordered the Brookes' land to the north were one of the older established Suffolk gentry families. They were making a fortune exporting cheese to Calais, Boulogne and to the king's army at Berwick-upon-Tweed. Many such families had 'close commercial connections'; Reyce commented on the way the Suffolk gentry usually traded openly at markets and fairs.

Alderman Brooke's main business interests remained in London, he was one of the wealthiest men in the Grocers' Company, itself one of the wealthiest guilds. However, he was probably not in the top league of rich Suffolk gentry, but he must have had sufficient funds to have been accepted into the middle or even the upper ranks, those with an annual income of about £2,000 per annum or more. His riches were not flaunted; certainly the Suffolk estates were well within his means. His sole intention was to bestow landed wealth on his son, which he did in 1597; his purchase of the Yoxford estate, coincided with his son's marriage. In this litigious age, knowledge of his great wealth may have made his neighbours nervous. When he died in 1601 he was buried in St Mary Woolchurch, next to what is now the Mansion House, his heart was truly in the City.

It was his son, Robert Brooke and his young bride Joan, who benefited from Alderman Brooke's wealth. Robert was 26 and Joan was just 17 when they married and took up residence in Suffolk. After his father's death Robert's mother Ursula joined them at Cockfield Hall which then became their country seat in Suffolk. The Yoxford house is said to date from 1613; the date appears in a dining room window. This would correspond with the Brooke's Jacobean additions. However, there is much evidence of earlier Tudor work, including a gatehouse belonging to the Hopton era. So young Robert and Joan made changes and spent money updating their new residence; the sort of thing fashionable young couples with money like to do. There is evidence that Westwood Lodge was also undergoing a revamp at this time, but of a different kind.

FIGURE 34.
Cockfield Hall
Yoxford: detail from
Copinger's Manors of
Suffolk showing the
Brookes' country
house as it was before
nineteenth-century
alterations.
COURTESY OF SUFFOLK
COUNTY RECORD OFFICE

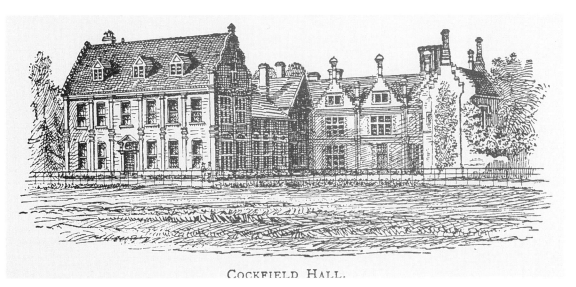

COCKFIELD HALL.

FIGURE 35.
The bell of 1601
which probably hung
in the cupola tower
over Westwood
Lodge. This was the
year of Alderman
Brooke's death. The
initials 'FH' and 'TC'
which also appear on
the bell do not seem
to relate to either the
Hopton or the Brooke
families.

In 1602, immediately after the death of his father, Westwood Lodge park was leased out for 21 years. The park was said to contain 240 acres with seven hundred trees, it was clearly of some significance because it was large enough to be marked on Saxton's map of Suffolk in 1575. The park was leased with all the enclosed ground lying in Blythburgh, Walberswick and Westleton, its out-buildings, orchards and gardens. The Brookes reserved the right to hunt with hawks in the park, but the lessee was allowed to kill all the deer. Provision was made for building repairs and, at first the Brookes came and stayed in the Lodge as 'resorters', but twelve years later in 1614, the house was let.

FIGURE 36.
The view south from the top of Westwood Lodge tower. Here Sir Robert Brooke would have been able to survey all that he owned and could see any tenants' cattle straying into his sheepwalks and marshes. He would have looked out over his formal gardens in the foreground, with Paul's Fen, in the distance on the right and the area of 'Bloody Marsh' to the left.

Francis Colbie of Kenton purchased the lease of the park and the enclosed land for £198 – a tidy sum – so cash was generated to make improvements at Cockfield Hall.

At the same time Robert was getting involved in local politics. He had begun a career in law, being registered at Gray's Inn, London, on 9 February 1593, aged 22. This career was probably curtailed when he moved to Suffolk four years later, but not his interest in litigation, which he persued avidly for the rest of his life. His library included 'law books' and books in Latin, which be bequeathed to his son, John in 1645. So some formal education seems likely. He was knighted in 1608, itself a formal recognition of gentry status, and became High Sheriff of Suffolk in 1614 at the age of 43. Being sheriff was an unpopular job, which most Justices of the Peace tried to avoid if they could, but there is no indication that he encountered any problems. In 1618 his first wife, Joan, died and was buried in Yoxford church. Within two years he had married a girl thirty years his junior, Elizabeth Culpepper, a cousin of his first wife. His suit was accepted because he was: 'a person of good estate and virtue' – this was a carefully worded statement made by a branch of his first wife's family, we must not take it as an unbiased character reference. For a time his new bride remained under the protection of his ex-mother-in-law, Lady Weld. Not a happy arrangement perhaps, since they lived as 'boarders', but children soon followed. Three of their surviving offspring were baptised in London, so much of their time must have been spent in the City. When they eventually settled at Cockfield Hall, Sir Robert's mother returned to London. One wonders why; the young Elizabeth's character may offer an explanation.

Of all the Brooke family Elizabeth Culpepper is the most interesting and remarkable. The Reverend Nathaniel Parkhurst gives us a flowing panegyric sermon on Elizabeth, from which she emerges as something of a religious fanatic; not an easy person to live with perhaps. Highly intelligent, a great

FIGURE 37. The new south entrance porch of Westwood Lodge reconstructed by Mr E. T. Webster.

reader of books, philosophical, puritanical, deeply religious and concerned for her children and household, she was constantly writing. Two substantial works survive and many other smaller manuscripts, including herbal and medicinal recipes in her own hand among the Cockfield Hall papers. 'What a Christian must believe and practice' and 'Collections, Observations, Experiences and Rules' were not published in her lifetime, but extracts were published at the time of her death and again in the nineteenth century. Delivered from his

FIGURE 38. Westwood Lodge gable detail. The house has been expertly reconstructed from archaeological evidence by Mr Webster using fragments recovered from behind a nineteenth-century facade. Sir Robert Brooke would have no difficulty in recognising the house he once lived in, which was built by the Hopton family in the sixteenth century.

Yoxford pulpit at her funeral in 1683, Parkhurst's sermon was published the following year together with Elizabeth's portrait (Plate VI), her biography and extracts from the writings.

The fact that Sir Robert chose to marry such an intelligent and cultured person, albeit many years younger than himself, and – bearing in mind attitudes towards women of the day – that he allowed her to study and philosophise openly, suggests that he was not a philistine and not unsympathetic to her views. In his will of 1645 he left her his books in English, reserving for his son those in Latin. Thus she may not have had a formal classical tutoring. She brought considerable wealth of her own to the marriage and purchased quantities of furniture and silver plate which was listed separately when her husband died.

There are indications that she had an iron will; attached to a lease of a farm in Yoxford, granted to one of her husband's trusted servants, is a note in Elizabeth's hand to say that it was granted: 'upon condition that Henry Woodward and Sarah his wife carry themselves pleasingly towards me, this shall stand, otherwise I will revoke it'. Her anxiety may have been prompted by a clause in her husband's will that required Henry and Sarah to get married before they could inherit his apparel. This was the same man described by John Barwick as: 'not worth a sucking lamb'; perhaps Barwick and Elizabeth shared the same opinion of the unfortunate Woodward. Religion may have become a refuge for her in the long years after her husband's death. Aged 82, deaf, harassed by law suits, Elizabeth outlived all but one of her children; she must have been a tragic figure living out her last lonely years in Cockfield Hall. She deserves more historical attention than this brief summary will allow.

Elizabeth's funeral was a spectacular affair costing £128 15s. and Parkhurst's panegyric was intended to match it. Nevertheless, this was the woman who provided morning and evening prayers with readings from the scriptures for her servants. After Parkhurst had been given the living of Yoxford, a: 'Grave Divine' visited her household once a fortnight to perform the office of Catechist, expounding the 'Principles of Religion' and examining the servants ...' a task Parkhurst himself had performed when he was her chaplain. The accounts detail a bible given to her gardener and purgative beer to her servants at springtime. This uncomfortable insight into a Suffolk puritan gentry household contrasts with any vision we may have of the Brookes as rogue squires. Elizabeth's style of religion as indicated by Parkhurst was an enthusiastic low-church Anglican, much in line with other leading puritan families in Suffolk. In her early years she was 'encouraged' by some

FIGURE 39. Portrait of Lady Elizabeth Brooke (née Culpepper), reproduced from Nathaniel Parkhurst's book of 1684.
COURTESY OF SUFFOLK COUNTY RECORD OFFICE AND THE BRITISH LIBRARY

divines who were non-conformists, and Parkhurst says that she: 'relieved many sober non-conformists with great bounty, and most earnestly desired to have them legally settled in a public ministry'.

Was she privately bank-rolling some non-conformist ministers before the Declaration of Indulgence in 1672? Parkhurst hints that not only was her charity very great, giving the tithes that she received, 'to him that had the care of souls' reserving only a small sum for church repairs, but that she gave away a large part of her fortune to, 'encourage the Ministry and to relieve the Indigent.' She entertained eminent divines such as Dr Sibs, Master of King's Hall, Cambridge. Sibs, who frequented her house at Langley, was also preacher to the 'Honorable Society of Grayes-Inn', London, where her husband had been a law student. Clearly her religious views tied in with her husbands network of friends. She also entertained the Bishop of Norwich, Dr Edward Reynolds, which would seem to suggest that she was Episcopalian, rather than Presbyterian in her views. She had probably moderated her puritanical stance in later years, as many people did.

She is said to have fasted in her closet during the execution of Charles I, 'preparing for the horrid murther of that excellent prince'. Becker hinted that Elizabeth was a closet royalist. But given her strong religious beliefs, her support for non-conformity before the Civil War, and her husband's position among the puritan gentry of the county, Elizabeth was almost certainly not a royalist; like so many at the time, she disapproved of the king's execution and was fearful of being party to the sin of regicide. Parkhurst was also being more politically correct than historically accurate, attempting to please his funeral audience of 1683 who would have come to appreciate Charles I as king, *and* martyr. He therefore presents Elizabeth as a moderate rather than a radical puritan, which she had been in her youth.

There is no doubt that the Brookes became inextricably involved with the puritan county gentry, thereby bolstering a healthy majority in East Suffolk with similar views. Sir Robert Brooke was MP for Dunwich in 1623/24, 1625, and 1628–29. He was not a member in 1626, nor was he in the Long Parliament of 1640, being then aged 69. Henry Coke and Anthony Bedingfield replaced him that year. Dunwich, of course, was a 'rotten' borough, but its population in the early seventeenth century was still substantial and although he owned much of the surrounding land, and many of the voters were his tenants it was unlikely to have been wholly in his 'pocket' in the way it might have been a century later. To some extent he must have wooed his electorate and won local support. He was not a particularly active member of parliament; this was before the days of organised political parties. Being an MP tells us more about his landed status and his concern to protect his own interests than it does about any political affiliation. Nevertheless, Sir Robert was a player in that 'tangled web of local custom, feud and prejudice' that passed for county politics.

Sir Robert was 'infirm of body' in May 1645. He died on 10 July 1646 at the ripe old age of 75, having enjoyed a long and successful life. If his will is

anything to go by, his religious views were moderately puritan, placing: 'my
soule into the hands of Almighty God hoping to be saved by the death and
passion of Jesus Christ etc.' It is a long will and this religious preamble is
relatively brief. To his 'wellbeloved' wife and sole executor Elizabeth he left
all his personal jewelry, gold chains, two coaches and four coach horses, all his
furniture etc. His books in Latin and his law books he left to his eldest son John.
In fact he had settled a substantial part of his estates on his eldest son John and
his daughter-in-law Jane before the will was written. 'Fortune' and 'Nature'
had smiled upon him as they had upon Fielding's 'Squire Allworthy'.

True there had been unhappiness, the loss of his first wife and at least one
child. There had also been endless legal battles. Nevertheless, blessed with
inherited wealth, two apparently happy marriages and four surviving children,
most of them well married, Sir Robert might truly be compared with Fielding's
archetypal 'good' squire, rather than the 'rogue' we are led to believe by some
of his unhappy tenants at Walberswick. If we knew no more about him than
the standard received history of the squirearchy that would be expected to
survive, and if we did not have the church wardens' accounts and John
Barwick's stories about him, we would judge him no better and no worse
than any of his contemporaries.

Both father and son were involved in the Civil War. His second son, Robert,
was one of the four deputy lieutenants of the county in 1642/43 who initiated
payment to troops mustered at Blythburgh. His fellow deputy lieutenants were
John Wentworth, from Somerleyton, Nathaniel Bacon, from Friston and
Robert Brewster from Wrentham. These were local men, staunch members of
the County Committee and part of the Eastern Association during the Civil
War. This is the clearest indication of the Brookes' acceptance into the Suffolk
puritan gentry and tells us where their loyalties lay during the War. Everitt
has commented on the stability of membership of the Suffolk Committee
compared with some other counties. The Wentworths, Playters, Rouses, Bacons
and Brookes, all from East Suffolk, survived as committee members beyond
1650.

Compared with the other committee men in their circle the Brookes were
relative newcomers, so their acceptance is interesting. Reyce talks about the
'interlacing' of gentry houses by marriage: 'much used for the strengthening
of families thereby.' Elizabeth, Robert Brooke's sister, married Thomas Bacon
of Friston. The Bacons were a leading and particularly active puritan gentry
family. Francis Bacon of Ipswich and Nathaniel Bacon of Friston sat on the
Committee for Scandalous Ministers which met several times at Yoxford, and
probably at Cockfield Hall in 1644–46. Another daughter, possibly Martha,
had married Robert Brewster of Wrentham, another strongly puritan East
Suffolk gentry family. Robert's brother John Brooke, the eldest son, made an
even better match, marrying Jane, the daughter of Sir Nathaniel Barnardiston
of Kedington, from the West Suffolk–Essex border. The Barnardistons were
by far the wealthiest puritan family in the shire with an estimated annual
income of £4,000. It is difficult to assess the exact wealth of the Brookes

without further research because they owned extensive estates in Hertfordshire, Essex and London as well as their Suffolk lands. Jane was a brilliant catch for the Brooke family and if this match demonstrates anything it is that as successful players in this 'exclusive country club' and marrying into the top draw of the county gentry, they must have had an annual income of nothing less than £2,500–£3,000 a year. It is likely the Brookes had this and more; it was money that made them socially acceptable, even if they were the 'grandsons of aldermen'.

Entertainment was the name of the game and Cockfield Hall, like the best gentry houses, was designed for it, being stylishly equipped with stables, gardens and a beautiful park clearly visible from the new London road. Much of the furniture, jewelry and silver plate in the house is listed in Sir Robert's will of 1645. Some pieces such as the 'blue furniture in my new parlor' and an empty cabinet given to his daughter-in-law Jane, were clearly new and probably very fashionable. Reyce says the gentry were much in the habit of visiting one another, while gambling at this period was endemic. The house was within ten to fifteen miles of many of the older puritan gentry family homes; a comfortable ride away (Map 4). Their guest list might have included the Norths at Laxfield and Benacre, the Heveninghams near Halesworth, the Brewsters at Wrentham, the Rous's at Henham, the Glemhams from Glemham Hall near Saxmundham and the Playters at Sotterley – these families have been described as the 'backbone of the County Committee' (Map 4).

So did the Brookes get on well with all their gentry neighbours? It would be surprising if they did. The Brookes got involved in many minor land disputes, mostly arising out of their purchase of the Blythburgh estate from the Hoptons. As we have seen in previous chapters, a thorn in Sir Robert's side was Henry Coke of Thorington Hall, the next big house up the London road (Map 4). The Coke family were among the most influential of the East Anglian gentry; they seem to have had many branches. Most influential was Sir Edward Coke, the famous lawyer, writer and originator of law reports. Although a Norfolk family, Coke tentacles spread into Suffolk and their estate at Thorington bordered the Brookes' lands at Hinton.

Sir Robert Brooke complained to Sir William Denny: 'that Sir Edward Coke haveing got a lease which the pryor granted to Richard Freeston (four score acres) before the disolution [of Blythburgh priory]; of the mannor of Hinton and the demenes thereof have gotten into his hands all the evidence concerning the same and kept the courts and hath created and granted by coppie the pryors haugh either he or his prediccessors and more they doe maintain and keep it as coppiehould and doe keepe away all my courtroules and other evidences which may manifest my title concerning the same.' Sir Edward Coke had purchased the lease from the Hoptons, but when the lease expired in 1635, it took twenty-three years of legal battle for the Brookes to get compensation for loss of their land.

The case involved two other neighbours, Edmund Stubbs a doctor of divinity and John Pepes who were under-tenants at Hinton. In Westleton Sir Edward

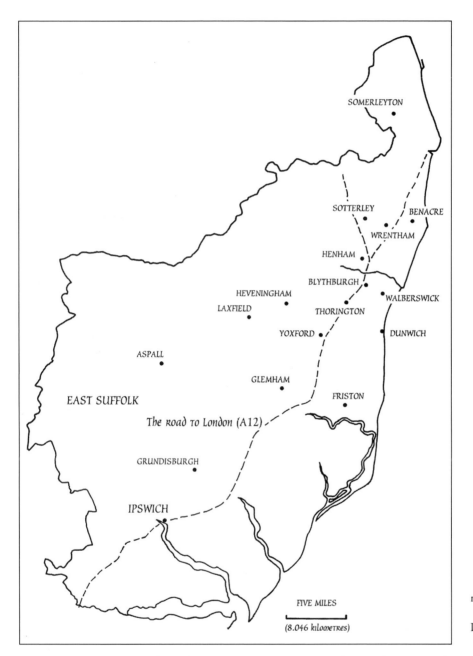

MAP 4. Places mentioned in Chapter 8 in relation to the London road (modern A12 and A145).

had also obstructed Brooke's claim to other lands: 'I have bin without remedy because I could never gitt any survayer to goe upon the lands to lay them out.' The Brookes used the most up-to-date methods of survey and mapping. For Alderman Brooke, nothing but the best would do; he commissioned Christopher Saxton to survey the Hinton Hall estate in 1594. Saxton's written survey and notes still survive, although only copies of the map now survive. In Westleton Sir Robert's claims remained frustrated. The Cokes were not primarily a Suffolk family so they were unlikely to damage his position among the county puritan gentry. Sir Edward Coke died in 1634, but the legal wrangling continued. Sir Robert won a court order from the Privy Council at Whitehall demanding that Henry Coke, Edmund Stubbs and John Pepes obey an order made against them in February 1638. But even this was ignored and the case rumbled on. Eventually in January 1658 Henry Coke agreed to pay £180 to settle the matter; honour was satisfied it seems.

Sir Robert Coke and Henry Coke were Sir Robert's fellow JPs and there is no doubt that they were hostile towards him. When Brooke had appropriated Walberswick quay in 1637/38, Henry Coke and others challenged him in the Court of Exchequer. The matter was not resolved until May 1639, when the petition of William Turrould was heard in the Court of the Inner Star Chamber, only then did Brooke give up his claim to the quay in the light of evidence from the town books (Document 10). At the same time there was another row involving the Court of Wards and Liveries. In January 1637, William Shrimpton, John Morris, Henry Jenney and Daniell Wicherley were ordered to attend the court at Westminster to answer questions by John Brooke in a suit against him by George Colt.

The Brookes seem to have taken all their legal battles as a matter of course and it has to be said that in most of them they were successful, even if it took many years of litigation. The legal training that Sir Robert gained in his early twenties, in London, had served him well. Such endless litigation was common in the sixteenth and seventeenth century; the established gentry frequently complained languidly of 'wearisome lawsuits' which dragged them to the fleshpots of London: not so the Brookes. The energy with which they pursued litigation was remarkable and in some respects it distinguishes them from the established gentry in the 'County Community'. Perhaps they lacked the social skills which would have enabled them to solve disputes with their country neighbours in other ways. There is also a certain paranoia in their litigiousness that suggests a basic underlying insecurity – itself, one suspects, the driving force behind their anxiety to succeed. It is recognisable also in their desire to marry into the wealthiest county families – to be seen as top dogs. There is a pushiness about them that is interesting: they represent the second generation of vigorous new blood pumping through the veins of the county gentry, but for 'blood', read 'money', a copious supply of which, it seems, made them readily acceptable to the older gentry familes.

Robert Brooke's name appears early in the Suffolk Committee Book during the Civil War, suggesting that he was an active member, but it is less frequent later on. Like many of his contemporaries he probably became disillusioned with the puritan Commonwealth. His name appears with his brother John among the list of signatories to the 1660 Ipswich petition demanding a free and full parliament. He was knighted for being a member of the parliament which brought about the restoration of the monarchy. Robert never inherited his father's estates and may have spent some time abroad during the later part of the Commonwealth. He was drowned while travelling in France in 1669. His elder brother John had inherited on his father's death in 1646, but John was short-lived and it was his wife Jane who ended up running the estate until she married again in 1660, to her sister's widower, William Blois of Grundisburgh. Thus the estates came into the Blois family.

All in all there is little to distinguish the Brookes from other higher-ranking gentry families in seventeenth-century Suffolk. They had become seamlessly integrated, socially, politically and economically; so much so, that if one was to look for a well documented typical puritan family among the ruling county oligarchy of the seventeenth century they would be an obvious choice. Apart from the younger Robert Brooke's participation in the restoration of monarchy, they were staunchly puritan. But people's political opinions changed then as they do now in the light of experience, particularly from generation to generation. The Brookes simply followed political, national and local trends.

Of course, there were those local families who fell foul of the County Committee for their Catholic sympathies and were cruelly punished for it, sometimes unfairly. Such were the Scriveners of Sibton who were too closely related to Matthew Scrivener the author of works on high Anglican teaching. John Scrivener acted as Sir Robert Brooke's lawyer in 1636, taking down statements against William Turrould. Curiously, he was one of the magistrates to whom Turrould wanted to have his petition referred, but all the others were committee men. Either Turrould was politically naive, or he knew more than we do. Such contradictions render political interpretations at a local level meaningless (Document 4).

Using their London money the Brookes had, from the start, made changes to the old Hopton estates – by building flood-walls and enclosing marshes from the sea and by enclosing the heath-grounds. They had allowed the magnificent park at Westwood to be dispoiled for economic gain. They split up their estates and leased them out to farmers who cared little for the old order, and who over-grazed the heaths with large flocks of sheep effectively excluding the tenants' beasts. They charged higher rents and thought nothing of impounding tenants' cattle when they strayed into the farmers' corn. The Brookes were simply doing what many other large landowners were doing in the mid-seventeenth century. They instigated an early agricultural revolution on their own estates which their customary tenants did not like. This revolution was enduring; not only did it create rich farmland, Westwood Lodge was regarded as one of the best farms in the county one hundred years later, but also it shaped the modern landscape in so far as there have been very few subsequent changes caused by later agricultural developments. The Brookes represent a new order, their approach to agriculture was revolutionary and uncompromising to the time-honoured ways of medieval communal grazing.

The Brookes were not 'rogue' gentry, at least they were no more 'rogues' than any of their contemporaries. They were, however, more than usually litigious, also, as part of a conscious policy of development and enclosure the Brookes created a divided community. Their newer tenant farmers on the enclosed grounds, such as Richard and William Wincopp and Thomas Palmer, were their staunch supporters, but we hear little or nothing about them. As far as the older Walberswick customary tenants were concerned, the Brookes were their sworn enemies, and through the legal documents they castigated their landlords, sometimes with good reason. Their rents were increased, their common rights were restricted and the town facilities, such as the quay and marshes were commandeered and leased out. Their anguish echoes down the centuries through the bundles of legal documents they generated.

The Brookes were looking towards land improvement and increased profit, but economic efficiency can be easily misinterpreted as meanness and exploitation. In consequence their relationship with the customary tenants was, in modern jargon, disfunctional. For a long time the wound of disenchantment slowly festered until 1637 when it became a suppurating sore of litigation, erupting violently in 1644 with the battle of Bloody Marsh, then slowly beginning to heal after the death of John Brooke the 'Great Troubler' in 1652. In the process the community of Walberswick was riven apart and radically changed, greatly diminished in size and wealth, and embittered emotionally, as it was dragged kicking and screaming into the early-modern world.

Rogues and neighbours: chapter notes

Macaulay's comments on the English esquire of the seventeenth century have a strongly Victorian flavour which does not bear historical scrutiny, nevertheless, its snobbish reference to the sons of aldermen is apposite: Macaulay 1848, p. 58. The ghost of John Brooke see: Becker 1935, p. 70; Mills West 1984. Meade Falkner published *Moonfleet* in 1898. Henry Fielding published *Joseph Andrews* in 1742. For the story of the Murgatroyds see the National Trust guide book to East Ridlesden Hall (1982). Gough's history of Myddle: Hey 1974. Landbeach in 1549: Ravensdale 1968. For the *Breviary of Suffolk* see: Hervey 1902. Dream dialogue: Becker 1935, p. 79. Antiquarian sources concerning the Brookes' ancestry: Copinger, Vol 1, p. 63; Gardner p. 141. Everitt 1960, p. 17 summarises the trade involvement of the Suffolk gentry. Cockfield Hall: Pevsner & Radcliffe 1975; Sandon 1977, p. 49. For the 'rebuilding' of Tudor and Stuart houses: Platt 1994. Will of Sir Robert Brooke: SRO, HA30: 50/22/27.3. For Gray's Inn Register: Foster 1889. Life of Elizabeth Culpepper/Lady Brooke: Parkhurst 1684; Dictionary of National Biography (DNB), 1917, II, 1328. References to Elizabeth's own personal effects: SRO, HA30: 50/22/27.3 [11]. Documents relating to Henry and Sarah Woodward, apart from Sir Robert Brookes will mentioned above: SRO, HA30: 50/22/1.28 [5]; CWA, fol. 147. For the reference to public ministry see: Parkhurst 1684, p. 72 and to Dr Sibs and Dr Reynolds, pp. 56–7. For the background to the puritan gentry in Suffolk: Everitt 1960; Blackwood 1997. The degree to which seventeeth-century members of Parliament canvassed their electorate is discussed in: Underdown 1985, p. 123. Quotation concerning county politics: Everitt 1960, p. 13. Quotation from Reyce: Hervey 1902, pp. 59–60. Committee for Scandalous Ministers meeting at Yoxford: Holmes 1920. Estimates of family incomes: Everitt 1960, p. 16. Sir Edward Coke: DNB, 1917, IV, 685–700. Complaint to Sir William Denny: SRO, HA30: 50/22/3.1 [2] and the litigation with Stubbs and Pepes in the same archive: [5–10, 13, 16–21, 37]. Saxton's notes, a photographic copy of the map and his survey of Hinton: SRO: HA30: 50/22/12.3. The Privy Council Order against Stubbs and Pepes: SRO, HA30: 50/22/3.1 [16] and the final settlement: SRO, HA30: 50/22/3.21 [11]. Exchequer proceedings: SRO,

HA30: 50/22/3.1 [51]. The George Colt suit: SRO, HA30: 50/22/3.1 [4]. For a discussion of the county community in Stuart historiography: Holmes 1997. References to the Brookes in the Committee Book during the Civil War: Everitt 1960, pp. 39, 127–8. Scrivener: DNB, 1917, XVII, pp. 1064–5; Everitt 1960, p. 27. For the background to the changing landscape of England during this period: Cantor 1987, pp. 44–8, 59.

Bloody Marsh

FIGURE 40.
Seal impression from the Blything Hundred court over which Sir Robert Brooke presided.
GARDNER 1754.

CHAPTER NINE

Farmers and factions

Lo! where the heath, with withering brake grown o'er,
Lends the light turf that warms the neighbouring poor;
from thence a length of burning sand appears,
Where the thin harvest waves its withered ears;
Rank weeds, that every art and care defy,
Reign o'er the land, and rob the blighted rye:
There thistles stretch their prickly arms afar,
And to the ragged infant threaten war,
There poppies nodding, mock the hope of toil;

Crabbe

In response to the order given by the Inner Star Chamber on 22 May 1639, the townsfolk of Walberswick brought an action against Sir Robert Brooke that summer in the Ipswich County Assizes. Edward Howlett, a tenant and inhabitant of Walberswick, was instrumental in bringing this action against Sir Robert. The Howletts had been established at Walberswick at least since 1582/83. Ironically, it was Nathaniel Howlet, almost certainly a descendant of Edward, who took over the tenancy of Westwood Lodge Farm in 1771 and who was commended by Arthur Young for then running: 'the finest farm in the county.' Howlett's case was heard before Sir Edward Littleton, a Shropshire man. Littleton was highly regarded as a judge, so much so, that not long afterwards, in January 1640, he was made Lord Chief Justice of the Common Pleas, and became a member of the Privy Council and a senior minister; a fact that was not to be lost on Sir Robert Brooke.

Howlett's challenge was a popular one, this time strongly supported by the townsfolk. The reason for the choice of Howlett could be no more than the fact that he was a substantial tenant of the manor, whereas William Turrould, being the miller, was a tenant of the town. Howlett lived in one of the larger houses in Walberswick, a point to which we will return later. Turrould may have been ruined financially by his three-year legal battle against Sir Robert. A large part of two summers had been wasted in London with his legal counsel attending court waiting to be heard. He does not appear among the list of Howlett's supporters, but he was still resident at Walberswick in 1644 and it is hard to imagine that he did not support Howlett's case. Turrould disappears from the next round of documents as the emphasis is placed on those tenants of Sir Robert who were deprived of grazing rights.

For the most part, we only have Sir Robert Brooke's version of the Howlett case; although biased, it is none the less revealing. Howlett accepted that Sir Robert was the rightful lord of the manor and that he was one of his tenants. His complaint was that Sir Robert had prevented him from grazing his great beasts on the common and heath grounds. He maintained that: 'The tenants of the said mannor inhabiting within the town of Walberswick have used to feed their great cattle of all sorts at all times in the year in a marsh called Pauls fenne conteyning 60 acres lying in Walberswick and in like manner have fed their cattle in a marsh called East Marsh conteyninge 20 acres lying

FIGURE 41. The wind-pump on Westwood marshes in *c.* 1930, before it was burnt out in the second World War. Gardner (p. 111) mentions a wind pump at 'Cuckolds Point' near here in 1743. East Hill and East Walks can be seen clearly in the distance.
G.F. BLOMFIELD

92

in Walberswick'. The names of eleven venerable witnesses appear with a folio reference to the town books together with their ages: thus Christopher Frost was aged 63 years; Anthony Gester 50 years etc. (Document 6).

Similarly, eight tenants could testify that Pauls fen lay in Walberswick and had, 'been taken in by the inhabitants when they have gone their perambulacon' – when they had beaten the bounds at Rogationtide (Document 16). True, Sir Robert's agents had enclosed Pauls Fen and East Marsh with a wall, but his father did not enclose them when he walled in other marshes from the sea. John Barwick and Edward Burford kept a town book, 'wherein the charges of making passages and draynes unto and in the said Paules fen are set downe to be done at the cost of the inhabitants.' For example, in 1588 when workmen were paid for 'trenchyn' in 'Palles fen'; indeed such references were sidelined to draw attention to them in the churchwardens' account book. This proved to be the most persuasive evidence. Both Barwick and Burford were to become key players in subsequent legal battles against Sir Robert. But where exactly was Pauls fen and East Marsh? Gardner marks 'Pauls fenn' on his copy of the Agas map of 1587 at the eastern end of the area which is now known as Westwood Marshes. This is almost certainly incorrect. Great Pauls Fen is marked on the Westleton Tithe map as a marsh enclosure in the area north of Foxburrow Walks. Also it is clear from other sources that the most contentious part of Pauls Fen lay directly south of Westwood Lodge in full view of the house (Map 1).

East Marsh lay below East Hill and East Walk in the area where Gardner marked 'Pauls fen'. This may explain Gardner's confusion over the location of Pauls Fen, for he must have felt compelled to place it in Walberswick. John Barwick claimed that East Marsh had previously been common marsh and was newly named East Marsh (Document 14). It is clear from early documents that Pauls Fen consisted of several pieces of land, which lay in the three parishes of Blythburgh, Walberswick and Westleton. The parish boundary at Pauls Fen was itself disputed between the two parties, hence the need to record the beating of the bounds in the town books. Brooke claimed that he had purchased, 'one other Marsh called Paules Fen contatneing 16 acres abbutt (as appeareth by the old tinderse) upon the marsh of the Lord of the said Manor which is called Paues Fen.' He also declared: 'that Paules fen lyeth in Blyburgh' (Document 5). The townsmen on the other hand were adamant that Pauls Fen lay in Walberswick, and, 'have been and yet is taken in by the inhabitants thereof when they have gone their perambulacon' (Document 6).

Several descriptions of the beating of the bounds appear among the churchwardens' accounts. In 1644, part of the journey was done by boat, but the perambulation of 1678, which is repeated by Gardner, is the most descriptive:

'From the channel cross the marshes, by the further end of Palls-Fenn, westward unto the Commons, where there were three crosses by the side of the marshes belonging to the house where now Goodman Hows the shepherd dwells; and from thence, back by the old bank, 'till we come right against the Park-House, going in at the great Garden Gates, and then right forth cross the Park, where there are several Trees marked with letters, and marks of our inhabitants, and thence to the Maple by the Park-Side, where there is a cross upon the common, and thence right cross to Deadmans-Cross, where boys heaved stones to the old heap, according to their old custom' (Document 16).

If we accept the evidence from the Westleton tithe map, the original parish boundary of Walberswick must have extended westward towards Hinton and

FIGURE 42. Fen Cottage, Blythburgh. This, or its predecessor, was where Goodman Howes the shepherd dwelt in 1644 when the townsfolk of Walberswick beat their parish bounds.

taken in part of Pauls Fen nearly to the modern B1125, while the modern parish boundary turns north before it gets to Pauls Fen. The older parish boundary was probably the one described by the townsfolk, whereas the modern boundary was the one favoured by the Brookes (Map 1). The 'crosses' mentioned in the bounds are cross-roads or cross-ways. The three crosses beside the marshes therefore refer to the cross-ways near New Delight Walks, where there are now two separate double-dweller cottages. The older one, called 'Fen Cottage', is probably the house of Goodman Hows the shepherd. The 'old bank' must be the prehistoric earthwork referred to in a sixteenth-century survey of Westleton as the 'Wenerdyke'. Returning eastwards along the line of this bank would bring the Rogation procession to a point opposite the 'Great Garden Gates of Westwood Lodge' (Document 16).

Howlett's witnesses declared that the tenants and inhabitants of Walberswick had enjoyed common of pasture for their great beasts on the 'great heath' called Walberswick Common at all times of the year, and that the heath extended from the 'townes end of Walberswick to Deadmans crosse' – so named after a suicide burial where the old road crossed the parish boundary (Documents 6 and 16). The tenants and the inhabitants had usually fed their sheep with a shepherd on the heath; the number of their sheep being three or four hundred. Eight tenants testified that when any of the heathland was ploughed and sown with 'corn' by the lord of the manor or his under-tenants, 'that great part of the said corn have been reaped and carried away by the commoners of Walberswick' (Document 6). They also caught rabbits using dogs and nets. Without attempting to describe the different areas of the heath, they acknowledged that the tenants of both Blythburgh and Walberswick had several different wastes where they each 'have common by themselves.' Thus they hinted that the picture was rather more complicated than their testimony suggested, certainly the rental of 1582/83 makes a distinction between four different types of heathground: the heath of the lord of the manor and the heath of the lord which was tenanted, the heath of the town of Walberswick and the copyhold heath of the town which was rented or tenanted: '*brueria nativum Villa de Walberswick*'.

In the churchwardens' account book there are several lengthy passages about the struggle for the commons written by John Barwick, some of which were repeated by Gardner. Barwick was a key Howlett supporter and 'penned' his story shortly before 1654, nearly fourteen years after Howlett's case was heard (Document 13). Although written in a rustic and anecdotal style, it is a graphic statement from the tenants' point of view. Needless to say, it is not without bias; at times it descends into gossip and innuendo. In places the order of events is not entirely accurate, but this can be ascertained by reference to other sources. The main drift of the gossip in Barwick's story concerns allegations of corruption and jury-rigging made against Sir Robert Brooke and his son John, which, it should be said, was probably not all one-sided. However, John Barwick said that Howlett produced twelve able witnesses, but after Judge Littleton had heard just six of them he decided in their favour: 'saying, that six was enoowe after he had seen the town aunchant bookes, which did playnly showe what cost the churchwardens had bine out in those days about the common, as in making way into Paules-Fenn, and allsoe on dickinge and drayninge of Paules-Fenn.' The town was thereby spared the expense of a long trial (Document 13).

So Edward Howlett won his case before Judge Littleton in the summer of 1639, and the townsmen of Walberswick won back their commons. The town quay had also been returned the previous year. The remaining bells of Walberswick church must surely have rung out that summer as never before. In gratitude, ten of his supporters: John Chester, Robert Bouton, Nicholas Girling, John Bowman, Edward Burford junior, 'old man' Henry Richardson, the redoubtable John Barwick, John Farercliffe, John Chapman and Robert Miller, together purchased Howlett's house in Walberswick and gave it to the town. Presumably this was to help him cover his costs. However, not everybody was happy with this idea: 'Edward Burford the elder being one of the cheefe townsmene for age and estate could not be persuaded to give any thing towards this pourchase soe that we 10 above did it with out his help.' Surprisingly the town was not united in its support for Howlett even in retrospect. Described as a 'Porch-house', the building stood at the junction of Fishersway and Maynfield, and was given to serve as a 'Town-house', in perpetual memory of Edward Howlett's service to the community. It survived until 1749 when it was consumed in the great fire of Walberswick witnessed by Gardner.

In the autumn both parties agreed an 'Accord' resulting from the trial. If real stories had a happy ending, here is where our story ought to end. But this was the seventeenth century in a country on the brink of civil war. The governance of the land was in turmoil; parliament had not sat for eleven years and a small elite group of the king's personal advisers was running affairs as they pleased. Anyone in personal contact with that elite group, or who knew the networks by which it might be influenced, was in a strong position to work the system in his favour. Sir Robert Brooke was just such a person. Aware that Judge Littleton was newly-made Lord Chief Justice of the Common Pleas, in the following year, Sir Robert being well versed in the law, appealed to the King's council at Whitehall of which now 'Lord' Littleton was a member, for a retrial (Document 12).

Sir Robert's appeal was heard at Whitehall on 19 August 1640. Around the council table sat the twelve most powerful and distinguished men in England: Archbishop Laud, the Lord Keeper, the Lord Treasurer, the Lord Chamberlain, the Earl of Dorset, the Earl of Holland, the Lord Lieutenant of Ireland, Lord Cottingham, Sir Thomas Rowe, the Secretary 'Mr' [Sir Francis] Windebank and last, but not least, Sir Edward Littleton. The basis of Brooke's appeal was simple and probably had some basis in truth: 'at the last sommer assizes for the said county of Suffolk: the said cause was tryed before Sr Edward Littleton Kt., Lord Chief Justice of the Common Pleas, a member of the board, which cause being popular the jury (being divers of them returned upon tales) gave a verdict against the petitioner [Brooke], contrary to the expectation of those that heard the evidence, which if the petitioner be concluded thereby, will prove very prejudiciall unto him' (Document 12).

In other words he was arguing that the jury had made a perverse decision against him. This, of course, is very different from the story 'penned' by John Barwick, who emphasised the over-riding decision of Judge Littleton, who was swayed not so much by the statements of the six town witnesses, as by the evidence from the town books. Now Brooke was, 'humbly praying their lordships to heare the opinion of the said Lord Chief Justice now present in Counsell.' It is impossible to know what happened between the autumn of 1639 and August 1640 to change Judge Littleton's mind. But support Sir Robert he did: 'The Lord Chief Justice now present in Counsell, expressing, that according to the business lay before him at the said asseze, he could not be of other opinion, but that it was requisite and fitting, the petitioner should have another tryall' (Document 12).

Littleton was a famous and distinguished lawyer; in the common pleas he had shown 'great ability', but in the Chancery he has been described as an 'indifferent judge' and in the Council he was 'out of his element'. He was said to be so disturbed by the unhappy state of the king's affairs that in 1641 he became ill. Whatever caused the reversal in Howlett's case, we shall never know; the decision of 1639 was stayed until such time as there could be another trial. There is no indication that the townsfolk of Walberswick were represented at that hearing in any way. Sir Robert Brooke must have walked out through Whitehall Palace Yard that August day with a smile on his face.

For less than two years after the summer of 1639, following the initial judgement of Littleton, the townsfolk enjoyed some use of their commons, but only with constant harassment from Sir Robert in the courts and his farmers on the ground. With Sir Robert's appeal in the autumn of 1640, the judgement was stayed until there could be another trial, and this generated an atmosphere of tension and uncertainty. Meanwhile, the churchwardens tried to re-assert ownership of Pauls Fen and East Marsh, but they did not meet with unanimous support. John Fayercliff and Edward Howlett went around the town asking for money to clear out the dikes and drains in Paules Fen: 'and we received from them to there power much willingness and cheerfulnesse. But from some that had most cattle there many more than any other [such] as John Cheshire never gave a penny yet I say we collected above 40s.'

Sir Robert did not give up on his litigation either, for in 1641 he sued Henry Richardson, John Fayercliffe, John Trappit and once again Edward Howlett, over their claims to common rights. Richardson had leased the quay of Walberswick in 1630, which had caused great antagonism in the community; then he had to forgo the lease when the quay was returned to the town in 1639. Having been a supporter of the squire, it seems that Richardson now found himself fighting Sir Robert in the courts. Henry was the grandson of Alexander Richardson who had bequeathed an almshouse called 'Popes' to the town in 1571. The Richardsons were an established and wealthy Walberswick family and, in spite of his involvement with Sir Robert over the lease of the quay in 1630, Henry's name appears frequently among the elders in the churchwardens' accounts both before and after that date. No doubt there were some members of the community, like Henry Richardson, whose loyalty was ambivalent until pushed into a corner over their grazing rights by Sir Robert.

At the same time John Barwick, Edward Howlett and others tried to re-establish their rights over Pauls Fen by putting cattle into the marsh, but their beasts were almost immediately impounded and driven into Westleton pound by Henry Woodward, one of Sir Robert's personal servants from Yoxford and a man called Read, one of Richard Wincopp's men; to John Barwick they were: 'a copple of creatures hardly worth a sucking lam.' The pound was less than four miles from Pauls Fen, but it was over 12 miles by road from Walberswick. Barwick claimed it was 20 miles, which is typical of his tendency to exaggerate. 'After they weare in the pound Sir Robert himself cam riding in his coach to the pound and gav order to the pounder that he should not leat them out to milk no give them any but let them starve if the owners fitch not a reprever ...'

One of Barwick's beasts was a 'lame cow' in calf. After 'nigh a week' the cow lost her calf and they were 'gatt out' in a sorry state. Barwick's story is not entirely clear, but it seems likely that he and others managed to break the animals out of the pound and they were sold for a lot less than they were originally worth. They tried in vain to bring a case against Read and Woodward; Read they manage to catch and fling into Blythburgh jail, but Woodward eluded them; in the end they gave up. John Barwick reckoned the whole episode cost him 50 shillings. In 1643, after repairing the bridges over the dikes, the towns folk again put their animals into Pauls Fen. But Sir Robert gave warning that he would impound them: 'then all people hasted out with there cattle: but my poore lam meare was the lag most ... I say she was forst into Bliborowe pound.' To cut a long story short, after three months in the pound John Barwick's mare died. Then: 'Sir Robert by violence with men and neasty dogges, day and night, kept from us Paules Fen a half year' (Documents 13 and 14). So the dispute slowly escalated into physical violence, but worse was yet to come.

Thomas Palmer, Sir Robert Brooke's 'great farmer', leased the ground nearest to the town, and before long he clashed with his fellow inhabitants. Palmer, like the Richardsons and the Howletts was from an established Walberswick peasant landholding family. There are indications that there was rivalry over land acquisition among the wealthier peasant families prior to the survey of 1582/83. There were certainly older rivalries between these families that went back further than the disputes of the seventeenth century. In 1643, Palmer planted a crop of rye in East Close, adjoining East Marsh – probably a winter-sown crop. The townsmen's sheep grazing in East Marsh crossed an overgrown dyke and fed into the rye. On 4 April 1644, perhaps in an act of revenge, he drove his flock of sheep down from East Hill into East Marsh. John Barwick describes how the men of the town rushed out and gathered on the hill above and looked down in disbelief; nothing could have been more provocative.

Barwick claimed that: 'as formerly this our common fenn which they have since called Est Mearch: this also by violence hath benn keept: having order to worrye the poore cattle with dogges: and to wound them and forse them into dikes to drown them.' These increasingly violent episodes marked a new level of tension between Sir Robert's farmers and the inhabitants of Walberswick. But was Palmer acting within his rights? After all Sir Robert's appeal had been upheld, the judgement which had returned the commons to the men of Walberswick had been stayed pending another trial. Was not Palmer simply acting for his master and in his own interests by returning the marsh back to the way it had been before?

Then, to prevent the townsmen returning: 'Sir Robert seat up a boarded howse neere Paules-fen, wherein he kept in men and great dogges day and night, to dispossess the poor inhabitants of there right of commonadge, and so did by violence take it away ...' (Document 13). In late April or early May of 1644, one of Brooke's henchmen, a nameless 'stout fellow', employed to keep the cattle off the marshes, came into the town of Walberswick and started quarrelling and fighting with the townsmen. In the ensuing afray this 'stout fellow' was mortally wounded. 'And Sir Robert soe followed it that 3 after were hanged, and [he] would have hanged more.' John Barwick claimed that the fen was thereafter known as 'Bloody Mearsh', but in other respects he makes little mention of the incident considering its seriousness. For some reason he does not tell us the names of the three townsmen who were hanged or the name of the stout man. These names must have been known to him, but for some reason he failed to mention them, nor do they appear in any other records. Perhaps the names were so well known at the time that they needed no mention, or perhaps John Barwick was more deeply involved in the incident than he cared to mention. If this was so it might explain why Brooke's bailiffs were so keen to find any excuse to arrest him (Documents 13 and 14).

The matter did not rest there. On 30 May the townsfolk set off to beat their bounds 'some in a boat' but were soon confronted with a 'boarded house' built for Sir Robert's henchmen and their 'great dogges'. Barwick also says that: 'About the 4th June in the night the new rayles and great newe poastes that were seat upon our wall at our common ffen, now of late called East Mearsh, these were sawen and cut into small peaces and thrown away into dikes about before they were seat twoe months' (Document 14). Thomas Palmer was also responsible, so it was claimed, for throwing back the 'manure' that had been scoured out of the dikes in Pauls Fen by the townsfolk (Document 13). So by these violent means the townsfolk were denied access to their marshes and common grazing rights. And that was the way the matter was left until 1646 when the old squire Sir Robert Brooke died and was succeeded by his son John, whose reputation was soon to far exceed his father's.

Such violence and loss of life was not unusual in seventeenth-century England, particularly in the poorer 'open', 'woodland' and 'fenland' rural communities. In the sixteenth and seventeenth century the homicide rate is estimated to have been in the region of 5–18 persons per 100,000 of the population. By comparison, in the period between 1930 and 1959 the rate was 0.4 persons. Living in the seventeeth century you were theoretically 10 to 15 times more likely to meet a violent death by manslaughter or murder than if you were living in the mid-twentieth century. Gough, in his history of Myddle in Shropshire could remember ten murders in the locality spread over a seventy-year period spanning the Civil War; many of these were the consequence of domestic violence, petty feuds and drunkenness. However, the four deaths at Walberswick probably aroused more concern at the time because they were the climax to a long-standing feud over enclosure. Becker, claiming Gardner as her source, says that Thomas Palmer was one of those killed and that the other three were hanged for his murder, but this is unlikely. Palmer was probably still alive in 1646 when his flock of sheep was purchased by John Brooke.

There is no reason to associate the battle of *Bloody Marsh* with any of the wider actions of the Civil War. However, the event itself is symptomatic of a general frustration with the inadequacies of the law and a breakdown in the relationships between different social groups in the local community. Such traumas could happen at any time, but the fact that it happened just two years into the Civil War could be significant. The process of law had failed the community; arguably it had failed Brooke as well as his tenants. Here, issues, local and national, which had failed to be resolved by an arcane and partisan legal system, boiled over into violence. On the one hand the microcosm of social and economic changes which led to the fight over *Bloody Marsh* can only be understood in relation to a wider world of national and governmental change, while on the other hand we have a better understanding of the national events by exploring the local circumstances that led up to *Bloody Marsh*.

Farmers and factions: chapter notes

George Crabbe, 'The Village': Book I, lines. 63–71. For Nathaniel Howlett see, Lawrence 1990, pp. 54–5. References to drainage: CWA, fols 19, 29, 59. Westleton tithe map and apportionment: SRO, FDA280/A1/1a&b (Pauls Fen, field 133) – this was probably an area of inter-commoning shared by the three parishes of Westleton, Blythburgh and Walberswick. The 'fossatum' called 'le wenerdyke' appears in: SRO, HA30: 50/22/27.6 and has been mapped: Warner 1982, p. 46. The modern road from Blythburgh to Walberswick running along the northern edge of Westwood Lodge Park was made in 1839: SRO, HA11/C9/60. Edward Burford's refusal to contribute towards the purchase of Howlett's house: CWA, fol. 146 (This is not mentioned by Gardner). Quotations on the character of Judge Littleton are taken from: DNB, 1917, XI, 1245–47. Fayercliffe and Howlett's collection: CWA, fol. 147. Evidence for further litigation: BL., Tanner 284,103; CWA, fol. 147r. There is a lengthy account by Barwick of the impounding of cattle in 1641 and events leading up to the *Bloody Marsh* incident: CWA, fols 147, 147r. Crime rates in seventeenth-century England: Cockburn 1977; Hey 1974. Sale of sheep by Palmer in 1646: SRO, HA30: 50/22/3.21 [8] also Becker, p. 68.

CHAPTER TEN

The final challenge

....

Then too I own, it grieves me to behold
Those ever virtuous, helpless now and old,
By all for care and industry approved,
For truth respected, and for temper loved;
And who, by sickness and misfortune tried,
Gave want its worth and poverty its pride:
I own it grieves me to behold them sent
From their old home, 'tis pain, 'tis punishment,
To leave each scene familiar, every face,
For a new people and a stranger race;

Crabbe

In October 1647, John Brooke eldest son and heir of Sir Robert sued John
Chapman, the brewer at Walberswick, for grazing his great beasts over the
800 acres of his enclosed land which lay between Black Hill and Deadman's
Cross. Brooke estimated his damages at £20. He was following exactly the
same course of legal action as that recommended by the Inner Star Chamber
to William Turrould and the tenants of Walberswick in 1639. Ten years of
legal battle and the aggression of Sir Robert's farmers had undoubtedly worn
down opposition in the town. This new case would test that opposition and,
if possible, break it by establishing a legal precedent: as John Barwick himself
commented, Chapman seldom had any beasts in the area in question, it was
not an area of grazing that was hotly contested (Document 13). Brooke's
argument as lord of the manor rested on his ownership of the ancient right
of Free Warren, which had been granted to the Lordship of Blythburgh in
about 1163–64, during the reign of Henry II.

Among the early documents there are few references to locations and areas,
but as the dispute wore on so there is increasing reference to specific places
where grazing rights might or might not apply. In 1640, when Sir Robert
Brooke successfully had the judgement of Lord Littleton overturned, there is
reference for the first time to the right of commonage in relation to 'three
several peices' (Document 12). From earlier sources we know that East Marsh
and Pauls Fen in the south of the parish constituted one of the areas and the
other was probably the marshes and walks adjoining Tinker's Farm to the
north. The third area comprised the heath grounds, some of which had been

converted into sheep walks, in the centre of the parish. This was where the legal battle was now focused (Map 1).

The 800 acres of pasture disputed between John Brooke and John Chapman in 1647 lay within the enclosures called the 'Walks' in Walberswick close to the boundary with Blythburgh. Sheep walks were large enclosures, sometimes containing several hundred acres, which were taken out of heathland in the sixteenth and seventeenth centuries. This 'walk' was described as 'alius heath' and had probably been open heathland before the acquisition of the manor by Alderman Brooke in 1592. The Walks were grazed by sheep and rabbits, and were periodically ploughed and sown with barley and rye. Never more than two crops were taken in successive years from the walks and they were then allowed to recover for several years as 'summerland', when they were opened up for rough grazing and sheep folds in preparation for further cultivation. The distinction between what was 'heath land' and 'sheep walk' was not always clear, the low banks topped with thorn bushes were not a barrier to hungry animals feeding on the thin vegetation.

John Brooke's lawyers noted that in the action brought against Sir Robert Brooke in 1639: 'att the triall Howlett provinge his right of common for great beastes upon the heath in question and in thother lands vidz in the marsh (which the inhabitants now enjoy) a verdict was given in general for Howlett against the plaintiff's father, though little or nothinge was proved for commoninge with sheep upon the said heath grounde.' There was little or no control over the tenants' animals turned out to graze on the tracts of heath, although the 'great beasts' were supposed to be tended by a 'follower' and the tenants' flock by a shepherd. It was now Squire John's intention to exclude the townsfolk from grazing the heath grounds on the Blythburgh side of the parish, in the area where he was expanding his sheep-walks, particularly in the area between Black Hill and the parish boundary at Deadman's Cross (Map 1). This amounted to nearly half of the remaining commonable area.

Brooke's terms of reference had to be precise. From a twelfth-century charter, of which he probably had a much later copy, he could establish: 'That the heath grounds in question are within the foresaid Charter Warren and Sheepes Walke belonging to the forsaid manor of Blythburgh wherein the inhabitants of Walberswick have libertie to feed their great beasts all over (except in the places sown with corne) and also their sheep upon some parts thereof with a follower (that is to say) as far as Black hill which is above a quarter of a myle from Walberswick townes end on the north syde thereof.'

From the old manorial documents he could easily prove that Walberswick was originally a hamlet of Blythburgh and therefore lay within his 'charter warren'. The name and precise location of Black Hill was more difficult however, but with help from five witnesses, two of whom were over 60 years of age, he argued that Black Hill: 'hath ben soe called for diverse yeares and that it is neere Walberswick wynd mill and that there is noe other hill neere the said mill.' However, a note underneath this statement says, '. . . four of the foresaid wittnesses did not know the said hill to be called by the name of Black hill, but will depose that the inhabitants of Walberswick have not fed there sheepe upon the heath further then the hill which is a little beyond Walberswick wynd mill which thother wittnesses call Black hill ...' These sheep walk enclosures had been in place for some ten years or so, since it was near Walberswick windmill that the miller, William Turrould, had attacked Sir Robert's men in 1636, and tried to stop them working on the new enclosure banks.

Another 16 witnesses could testify that the townsfolk had pastured their sheep on the heath, 'but did not feed them on the heath grounds ... between Blackhill and Deadmans Crosse.' In Howlett's case heard before Judge Little-ton, which had gone against Sir Robert, one of the witnesses named Anthony Gester had testified: 'that the inhabitants 50 yeares since did feed their sheepe upon Walberswick heath as far as Black hill, but deposeth noe further'. Gester was then a copyholder of the manor. These statements sound convincing, but in fact there is more than a degree of ambiguity. There was another 'Black Hill' which lay near Pauls Fen in Westleton and is clearly marked as such on the first series Ordnance Survey map. This Black Hill lies very close to the original Walberswick parish boundary, the one claimed by the townsmen. Before the boundary was changed, the townsfolk of Walberswick may very well have been entitled to graze their sheep *'as far as Black hill'* – that is to say Black Hill in Westleton, not Walberswick. There is no mention of a 'Black Hill' in the 1582/83 rental of Walberswick, but only the heathland closest to the town is described in it.

Did John Brooke knowingly using this ambiguous reference to confuse the issue, or was it a genuine mistake? Whatever the truth may be, the matter was brought before the assizes in the autumn of 1647, and Squire John carried the day. John Barwick was convinced of the Squire's underhand dealings, but only in respect of the 'fowle menes' he thought he had used to rig the jury; he gives us a complex story of gossip and innuendo about what he thought had actually happened (Document 13).

John Barwick says the case was heard by Judge Pheasant who, he claims, was a near kinsman of Lady Brooke. This was almost certainly Peter Pheasant, who was a Justice of the Common Pleas in 1643. His eldest son, Stephen Pheasant of Grays Inn, London, is mentioned as a 'cousin' in Sir Robert's will of 1645. Stephen was registered at Gray's Inn twice, in 1624 and 1629. Sir Robert himself had trained at Gray's Inn, where he was registered as the son and heir of Robert Brooke, Alderman of London, in 1593. One of his tutors would have been Peter Pheasant, then Reader of Gray's Inn and father of Stephen. Quite what the family connection was that led to Stephen being called 'cousin' remains uncertain. The Pheasants were an old legal family, fathers and sons being recorded in succession on the Gray's Inn register from 1561. The legal profession was just as incestuous in the seventeenth century as it was in later periods. No doubt Sir Robert maintained old friendships from his student days in London and used the network of fellow commoners from the same Inn of Court to assist him in his litigation, just as any lawyer might do today.

Barwick also claimed that the foreman of the jury, William Buckenhams, was another near kinsman and had worked for squire John as steward of his manorial courts. Bartholomew Bullen, one of the witnesses (who appears in the Brooke documents as Bartholomew Bunnell), was the son of Sir Robert Brooke's bailiff and was married to the sister of the jury foreman, William Buckenhams. The gossip was that this Bartholomew had come by a great deal of money, which he had then 'swaggered' away with his friends at Yoxford. Barwick implied that Bartholomew was the agent who had bribed the jury on Squire John's behalf. Another witness, an elderly steward of Yoxford, possibly William Stannard (aged 70 years), had likewise changed sides. Shortly before he died he admitted that he had given witness, 'as Robert Turner bad him.' A woman, possibly 'Mrs Shorte', testified on the Squire's behalf when she had previously testified for the town before Judge Littleton. The reason for her change of mind was that she now had a son who rented a farm from the squire. When William Emmingham, lawyer for the defence, held up her earlier statement, Judge Pheasant called her back, but she had gone and was found to be too sick to testify (Document 13).

More serious was the fact that John Chapman only produced six witnesses. Barwick says that this was a deliberate attempt to save expense; in the trial before Judge Littleton the case had been decided on six rather than twelve witnesses, which had saved time and expense. But Squire John now had twelve witnesses. Barwick insinuates that these twelve were being manipulated; apparently some of them had difficulty in speaking and so the judge allowed their counsel to 'help' them. In contrast, Barwick says that the six witnesses for the town, 'more worth than six score of the 12 that came against them,' in particular Robert Durrant of Coatby, Robert Hart of Henstead and 'old' Widow Wynes of Southwold; 'these six witnesses were soe forward in speaking, as this Judge Pheasant would not suffer them to spake up that which they had to spake.' Perhaps their forwardness so irritated the judge that he silenced them. However, in his summing up, Judge Pheasant pointed out to the jury that they should take into account the fact that there were just six witnesses on one side and twelve on the other, so the jury ruled against John Chapman (Document 13).

After the jury had given their verdict John Barwick tells us that: 'many of the jurymen seeing a man which they had thought had bine he that had promised them soe much a pece when they had doon; but when they see they weare mistaken, they slanke away; and that man, he revillinge them, telling them they weare roges, in wronginge of a poore towne, and nowe would you be rewarded, sayed the gallowes is there just reward. Thus the poore towne had the overthrowe, in overthowinge John Chapman' (Document 13). It is perfectly possible that Barwick was right and that a wicked fraud had been perpetrated. The jury, promised money if they would declare for the squire, did just that, only to find that not only was there no money forthcoming, but that they were likely to be prosecuted if it was found out that they had accepted the promise of a bribe. John Barwick was reluctant to name names, he clearly thought that Bartholomew Bunnell had become rich somehow as a result of the trial, but exactly who had paid him he would, or could not say.

In truth, it may have been difficult for John Chapman to get more than six witnesses to testify for him. Not only had many of the tenants departed since the celebrated case of Edward Howlett, but the enclosed ground in the parish was sprouting a new breed of tenant farmer and dependent farm labourers. These folk were not copyholders of the manor with any rights over land in the parish; ancient grazing rights meant very little to them compared with the prospect of a job and sufficient wages to keep them out of the poor house. Economic forces and changing social structures had won the day. Loyalties within the village had changed irredeemably in favour of the new tenant farmers, who had no interest in maintaining the old order and its antiquated medieval customs. The old community with its interdependent strip fields, labour services and common rights was fast disappearing.

But what of John Chapman, the brewer of Walberswick? Like the miller, William Turrould, he too had to face the hostility of certain factions within his community. In 1645, two years before he was prosecuted by John Brooke, he was indicted for selling beer at one shilling above the statutory price. But Chapman was of a more subtle breed than his counterpart the miller. At the Blythburgh quarter sessions, his council, Mr Beddingfield: 'produced an ordinance of Parliament that bruers were to pay two shillings the barrell for excise. This corte doth discharge the said indictment which said indictment being for selling beer att nine shillings the barrell.' So John Chapman got off the hook. One cannot help feeling that Squire John was up to the same old tricks as his father, using every means to discredit any person who opposed his enclosures.

When John Barwick 'penned' his version of events he too had been the victim of Brooke's 'wicked actions', which is another reason why we must treat his testimony with caution. Following the Chapman case of 1647, John Barwick himself was arrested for grazing with his great beasts in 100 acres of the squire's heath grounds. A trial was intended for that same winter: 'but it being too tedious a season, and hard journey, soe furre and so fowle for such aged lame people,' the trial was delayed until the mid-summer assizes of 1648. To achieve this John Chapman had to ride with all speed to London, on his brother Roger Barwick's mare, being then with us '... for he was to goe with speed, and soe did.' (Document 13). But in the mid-summer assizes there were legal objections to the case going forward and nothing was done. In the following year John Chapman was taken so ill that John Barwick had to stand in for him as churchwarden.

On 10 November 1651 John Barwick was arrested again by Bailiff Dowtey: 'he brought a tryall then agaist me because I would not give him his unjust demands: his Councellor and wittnesse did much fouly declare and largly swore agaist me. Yet through the goodness of God in giving wisdome to the goodman Judge Sir John, who found out the foul wittnesse bearer ... which if otherwise if he had not byn found out, I had byn then more than half undon. I define I may never forget it, that great wonderfull deliverance of myne.' Barwick recognised that in spite of this success he was up against a man with limitless resources: 'he know that we are a poore people and cannot appeare soe personally our ablenes cannot afford it, and he loves to take all advantages.' Barwick was himself in fighting spirit, but he was not as youthful as the young squire: 'for myne own part I am aged between 3 and 4 score yeares [60–80] and one unfitt to travell' (Document 16).

In spite of this, on 18 May of the following year, John Barwick issued his own summons against Squire John, the writ being served by the Sheriff, Mr Tuttle. However, ten days later, on 28 May John Barwick was arrested again, this time by bailiff Brytton on the orders of John Brooke (Document 13). The bailiff gave no reason, but Barwick knew full well why he had been arrested. A letter written by John Barwick, just three days before his arrest, to a 'Reverend Sir', probably the newly-appointed vicar of Blythburgh and Walberswick, Nathaniel Flowerdew, a local man (Document 16). Clearly the new vicar had expressed his concern and had written to the Squire about it. John Barwick thanked him for his: 'paynes in writeing to our adversary for us though he have been since much oppressing us with his whole fflock of sheep now just to our doores where he never had them went before ...' Barwick then took the opportunity to write a three-page letter detailing all his complaints against the Squire and a summary of events including the story of *Bloody Marsh* and how three of the townsmen had been executed. Of all the documents in the collection it is the most moving and pitiful account. We can imagine the vicar's dismay upon reading it; the likelihood is that he immediately showed it to his employer, Squire John Brooke, for this letter now survives among the Cockfield Hall papers. The Squire must have exploded with rage and immediately ordered Barwick's arrest.

In his letter John Barwick explains that James Reynolds, probably a lawyer, and another man had come to his house and that together they had prepared another petition against Squire John, which Reynolds would present: 'without any great charge.'

'I told him that we are soe poore; what little money our poore neighbours have towards it: yet myself being moved earnestly by some, my selfe with another man made a beginning and gave him his desire, and drew out a petition, and nothing but the truth what ever our adversary may say to the contrary: for indeed to bring his own ends about he cares not what himselfe and other [s] say and swear to it. I myself have much experience of him this way.' The letter goes on to describe a list of travesties and abuses that Squire John had perpetrated: his use of false witnesses, his leasing out of tithes and his failure to provide a separate minister for Walberswick, his abuse of the commons, the impounding and suffering of the townsmen's beasts, his appropriation of Pauls Fen, the ploughing of the upland commons, the abuse of gleaners and the bullying of tenants and labourers. It is indeed an agonising indictment.

Barwick says that his neighbours had been gleaning barley the previous year on some of the upland commons that had been ploughed. The squire had been in the field and seen them at work all day, but in the evening the squire himself came and carted away the corn his tenants had gleaned: 'though the poore people pleaded much, seeing they had soe spent the day, they told him if he had but held up his hand against them at the first they would have been gone.' Returning home they declared, 'they had this day gleaned for the esquire …' Gleaning is a good example of one of the ancient customary practices, so important to the poor, but open to abuse and not conducive to the new farming methods. Leasehold farmers could not afford to lose valuable grain from the harvest.

Arthur Young, writing in 1771, says that although the custom was very ancient, 'the poor have no right to glean … the abuse of gleaning, in many places, is so great as deservedly to be ranked among the greatest evils the farmer undergoes: the poor glean among the sheaves, and too often *from* them, in so notorious a manner, that complaints of it are innumerable.' He advocated strict rules to be observed, the crop should first be entirely cleared and no

FIGURE 43. Bundles of cut reeds for thatching from Westwood Marshes. Before the second world war these were prime grazing marshes maintained for cattle and sheep – reed cutting now helps to preserve an equally artificial habitat for the benefit of marshland bird species.

animals should be allowed in until gleaning was complete. Barwick never tells us that the townsfolk might be doing wrong or even that their actions were likely to be misinterpreted. He always stresses their complete innocence and there is no doubt that sometimes he 'protesteth too much'. In the 1650s the community at Walberswick was so impoverished that the temptation to pilfer food must have been overwhelming for some of the inhabitants. It is the over-reaction of the lord of the manor to these events that is revealing. This point is better illustrated in another story also told by Barwick.

An old man was ordered by the squire to cut reeds and bring them up to the manor house at Westwood Lodge: 'the rushes were cutt and carted most of them to this house whereon they were layed, the esquire hearing where the man cut them, which most of them upon the common waste salts where the esquire hath a waste salt rushy marsh whereon some were cut, the esquire appeared displeased, the poore man then left fetching of them soe that the remaynder they carryed away the esquire soe threatening this poore old man to have him into jayle that he went to him telling him that they were laid upon his own house at Walberswick.' The esquire refused to pay him for the work and continued to threaten him with jail for pilfering the reeds: 'he fell down on his knees humbly beseeching him yet no mercy could fynd, loth the old man was in his age to lye in the jayle, he got another man to go to the Esquire with the old man's wife and carryed with her 6 shillings, soe through much adoe they at last prevayled for the 6s. and 14 more there was paid to him before that midsommer according to agreement soe he had 20s.' (Document 14).

'This midsommer he brings a tryall against me [John Barwick] for goeing our bounds as we have gone allwayes before, and one we bring against him for our commons. Thus in the behalf of our town most of the seamen at sea, and 2 or 3 of our best well-wishers weak and sick in bed ...' (Document 14). These cases were brought once more to the mid-summer assizes of 1652, but again there were legal objections to their being heard, one of which may have been the sickness that was already affecting the town that May.

This was the year when both churchwardens, John Chapman and John Chettle, and at least one other parish officer, the 'wharfinger' died suddenly, due to an epidemic of some kind. But somehow John Barwick survived and again replaced John Chapman and continued to maintain the town books. Plague, which had struck towns in the west of England in 1650, spread to the rest of England and Ireland over the next three years. Seaports were particularly vulnerable to the plague carried by rat-infested ships. An epidemic disease, believed to have been influenza, also struck the seaside towns of Lancashire and Cheshire in 1651 causing alarm in Liverpool through into January 1652. It is possible therefore that one or other of these contagions reached Walberswick in the early summer of 1652. The exact number of deaths in the town remains a mystery since the burial registers do not begin until four years later, but given the impoverished state of the town the death rate may have been high.

The arrival of this epidemic probably overshadowed the increasingly acrimonious argument between Brooke and Barwick; the plague respected no one, those in service to the community, such as the churchwardens, were more likely than most to become infected. The local JPs must have been well aware of what was happening, yet Squire John Brooke still insisted on attending his court and he continued relentlessly with his legal actions against the town. However, fate, ever fickle, had its own course of action. John Brooke, like his father, was an energetic man; in September 1652 he had been busy supervising repairs to Westwood Lodge and had been out hunting. Getting up early one Saturday morning he complained of feeling unwell. He intended riding in his coach to Blythburgh that day where he would sit in the magistrates' court. His wife, Jane, and friends managed to dissuade him from going and he retired to bed. That night he died (Document 13). His death was probably natural; perhaps he succumbed to the same epidemic that had carried away the churchwardens of Walberswick. We will never know. There is no suggestion of foul play in any of the documents, which might suggest that the symptoms were well recognised.

John Brooke was buried at Yoxford and his feud with the townsfolk of Walberswick was buried with him. His wife, Jane, now a childless widow, left Westwood Lodge for good and went to live with her mother at Kedington. Under the 'virtuous Lady Brooke' the town again enjoyed its ancient privileges. Westwood Lodge was leased out the following year to Oliver Chattborn, but Jane kept the use of the principal rooms and her two maidservants maintained the house and kept the accounts in her absence. In May 1655 John Brooke's gravestone arrived at the quay of Walberswick on its way from London to Yoxford. John Barwick 'penned' his version of the feud sometime before January 1654, two years after Brooke and the churchwardens had died so suddenly, prompted perhaps by a sense of his own age and mortality and the desire to record the momentous events that had transformed his community.

So what sort of a man was John Barwick? The second son of Ewen and Margery Barwick, he had two brothers, Thomas and Roger and two sisters, Mary and Margaret. His father was an Ipswich fishmonger with business interests in Walberswick, owning jointly with his eldest son, Thomas, a lease on salt houses there and a 'joint stock in the trade of salt making.' To judge from his father's will the family were strongly puritan and John may well have had an education to match. Certainly he loved writing, a skill which enabled him to record the *Bloody Marsh* tragedy for posterity. He was himself a merchant, importing salt, coals and oil through Walberswick quay in the 1630s. His father had bought the tenement of William Clarke, carpenter, at Walberswick before 1623. The carpenters were mostly ship-builders or ship carpenters and traded in timber.

On a number of occasions John was elected as churchwarden, more often than not with his close friend John Chapman. His writing skills and religious opinions provided sound qualifications and presumably he was re-elected because he was trusted and worked hard. Salt production bridged both the fishing industry and the contentious coastal trades enabling him to understand different business interests and factions within the community. His willingness to openly challenge authority and speak his mind we may find endearing, but to the person being challenged he could be infuriating. There is an irritating fastidiousness about his writing and a tendency to mix and muddle different issues to suit his arguments.

He had a pedantic and unforgiving side to his character. We can enjoy

FIGURE 44.
Modern fence post on the heathland at Walberswick. The area is now jealously guarded from the public by English Nature. Where Sir Robert Brooke failed in his enclosures of the sixteenth century, modern authorities have succeeded in the name of 'conservation'.

some contradictions and exaggerations in his story – his outrage at the death of his own 'poore lame cow' in Blythburgh pound, matched only by his description of it as a 'good cow' when undervalued by the squire's bailiff. There can be no doubt that John Barwick was a good and honest man living in difficult times. In many respects he is the hero of *Bloody Marsh*, for without him the tale could never have been told. This book is therefore dedicated to him, his friends and his fellow churchwardens. We can only imagine his thoughts as he looked down in 1655, aged 'between 3 and 4 score years', on the black slab of marble resting on the quayside engraved with the name of his adversary, that 'great troubler', John Brooke Esq.

> The boast of heraldry, the pomp of power,
> And all that beauty, all that wealth e'er gave,
> Awaits alike the inevitable hour.
> The paths of glory lead but to the grave.
>
> *Gray*

The final challenge: chapter notes

George Crabbe, 'The Borough', Letter 18, lines 195–204. The reference to three pieces of common: SRO, HA30: 50/22/3.1 [26]. The passages referring back to Howlett's trial, charter warren and Black Hill: SRO, HA30: 50/22/3.1 [57]. The will of Sir Robert Brooke: SRO, HA30: 50/22/27.3 reads as follows: 'to my cousen Stephen Pheasant of Grayes Esq ...' For Gray's Inn Register: Foster 1889. John Chapman's illness and his deputies as church wardens: CWA, fol. 151. John Barwick's replacement of John Chapman as church warden: CWA, fol. 151. For the epidemic of possible influenza in 1652 see: Shrewsbury 1971, pp. 435–42. Thomas Barweck's will: Allen 1989, p. 485. The final extract is from Thomas Gray, 'Elegy Written in a Country Churchyard', verse 9.

Bibliography

M Allan, 1964. *The Tradescants: their plants, gardens, and museum, 1570–1662*.

M E Allen (ed.), 1989. *Wills of the Archdeaconry of Suffolk 1620–1624*. Suffolk Record Society (SRS) 31.

M E Allen (ed.), 1995. *Wills of the Archdeaconry of Suffolk 1625–1626*, SRS 37.

K J Allison, 1957. 'The Sheep-corn Husbandry of Norfolk in the Sixteenth and Seventeenth centuries', *Agricultural History Review* 5 (i), 12–31.

P Armstrong, 1975. *The Changing Landscape*, Lavenham.

B Ayers, 1994. *Norwich*, English Heritage, Batsford, London.

J & S Bacon, 1979. *Walberswick Suffolk*, Guidebook, Colchester.

M Bailey, 1990. 'Coastal Fishing off South East Suffolk in the Century after the Black Death' *Proceedings of the Suffolk Institute of Archaeology and Local History* (PSIA) 37 (ii), 102–14.

J Becker, 1935. *Blythburgh: essay on the Village of Blythburgh*, Halesworth.

M Beresford, 1961. 'Habitation and Improvement', in F J Fisher (ed.), *Essays in the Economic and Social History of Tudor and Stuart England*, Cambridge University Press, Cambridge.

J H Bettey, 1979. *Church & Community: The Parish Church in English Life*, Moonraker, Bradford-on-Avon.

J H Bettey, 1987. *Rural Life in Wessex 1500–1900*, Sutton, Stroud.

B G Blackwood, 1988. 'The Gentry of Suffolk during the Civil War', in D Dymond & E Martin (eds), *An Historical Atlas of Suffolk*, Suffolk County Council.

B G Blackwood, 1997. 'Parties and Issues in the Civil War in Lancashire and East Anglia', in R C Richardson (ed.), *The English Civil Wars: Local Aspects*, Sutton, Stroud.

L Cantor, 1987, *The Changing English Countryside 1400–1700*, Routledge, London.

C Chitty, 1950. 'Kessingland and Walberswick Church Towers' *PSIA* 25 (ii), 164–71.

J B Clare, 1903. *Curious Parish Records: Wenhaston and Bulcamp etc.*, Halesworth.

J S Cockburn, 1972. *A History of the English Assizes 1558–1714*, Cambridge Studies in Legal History, Cambridge University Press, Cambridge.

J S Cockburn (ed.), 1971. *Crime in England 1550–1800*, London.

N Comfort, 1994. *The Lost City of Dunwich*, Dalton, Lavenham.

W A Copinger, 1905–11. *The Manors of Suffolk*, 7 vols.

H C Derby, 1940. *The Drainage of the Fens*, Cambridge University Press, Cambridge.

D Defoe, 1991. *Tour through the Whole Island of Great Britain, 1724–27*, Yale University Press, Yale.

D Dymond & R Virgoe, 1986. 'The Reduced Population and Wealth of Early Fifteenth-Century Suffolk' *PSIA* 36 (ii), 73–100.

N Evans, 1984. 'Farming and Land-Holding in Wood-Pasture East Anglia 1550–1650' *PSIA* 35 (iv), 303–15.

Bibliography

N Evans, 1985. *The East Anglian Linen Industry: Rural Industry and Local Economy, 1500–1850*, Pasold Studies in Textile History 5, Gower, Aldershot.

E Everitt (ed.), 1960. *Suffolk and the Great Rebellion 1640–60*, Ipswich.

E Everitt, 1966. *The Community of Kent and the Great Rebellion 1640–1660*, Leicester University Press, Leicester

H Fielding, 1749. *Tom Jones*. (Dent, 1998, London).

H Fielding, 1742. *Joseph Andrews*. (Penguin Classics 1985, London).

J A Fitch, 1964. 'Some Ancient Suffolk Parochial Libraries', *PSIA* 30 (i), 44–87.

J Foster, 1889. *Gray's Inn: Admissions Registers 1521–1889*, Hansard, London.

Thomas Gardner, 1754. *An Historical Account of Dunwich … Blythburgh … Southwold … with remarks on some places contiguous thereto.* London.

F A Girling, 1961. 'Merchants' Marks in Suffolk' *PSIA* 29 (i), 103-32.

C Harper-Bill (ed.), 1980. *The Blythburgh Priory Cartulary*, SRS, Suffolk Charters 1 & 2. Woodbridge.

A Hassell Smith, 1989, 'Labourers in late sixteenth-century England: a case study from north Norfolk', *Continuity and Change* 4 (3), 367–94.

M Hawkins, 1973. 'The Government: Its Role And Its Aims', in C Russell (ed.), *The Origins of the English Civil War*, Macmillan, London.

S H A Hervey (ed.), 1905. *Suffolk in 1674: being the Hearth Tax Returns*, SRS, Suffolk Green Book 11.

S H A Hervey (ed.), 1909. *Suffolk in 1568*, SRS, Suffolk Green Book 12.

S H A Hervey (ed.), 1902. *Suffolk in the Seventeenth Century: A Breviary of Suffolk by Robert Reyce 1618*, London.

D Hey, 1974. *An English Rural Community: Myddle Under the Tudors and Stuarts*, London.

C Holmes (ed.), 1970. *The Suffolk Committees for Scandalous Ministers 1644–46*, SRS, 13.

C Holmes, 1974. *The Eastern Association in the English Civil War*, Cambridge University Press, Cambridge.

C Holmes, 1997. 'The County Community in Stuart Historiography', in R Cust & A Hughes (eds), *The English Civil War*, Arnold, London.

W G Hoskins, 1953–54. 'Harvest Fluctuations and English Economic History 1480–1619' *Agricultural History Review*, 2, 28–46.

A H Johnson, 1909. *The Disappearance of the Small Landowner*, Oxford.

H H Lamb, 1966. *The Changing Climate*, London.

R Lawrence, 1990. *Southwold River; Georgian Life in the Blyth Valley*, Wheaton, Exeter.

R W M Lewis, 1947. *Walberswick Churchwardens' Accounts AD 1450–1499*, London.

K Lindley, 1982. *Fenland Riots in the English Revolution*, Cambridge.

C P Lucas, 1905. *A Historical Geography of the British Colonies, Vol. 2 The West Indies*, Oxford, Clarendon Press (2nd edn).

Lord Macaulay, 1848. 'The state of England in 1685: Squires and Parsons', in his *History of England*, H Trevor-Roper (ed.), 1968, Penguin, London.

D MacCulloch, 1975. 'Ralph Agas' *PSIA* 33 (iii), 275–84.

J Middleton-Stewart, 1996. 'Down to the Sea in Ships: decline and fall of the Suffolk Coast', in C Rawcliffe, R Virgoe & R Wilson (eds), *Counties and Communities: Essays on East Anglian history presented to Hassell Smith*, Norwich.

H Mills West, 1984. *Ghosts of East Anglia*, Barbara Hopkinson Books.

J Morrill, 1984. 'The Religious Context of the English Civil War', *Transaction of the Royal Historical Society*, Fifth Series, Volume 34, 155–78 (reprinted in Cust & Hughes, 1997).

J Morrill & J Walter, 1997. 'Order and Disorder in the English Revolution' in R Cust & A Hughes (eds), *The English Civil War*, Arnold, London.

R B Outhwaite, 1982. *Inflation in Tudor and Stuart England*, second edition, Macmillan, London.

N Parkhurst, 1684. *Funeral Sermon on Lady Elizabeth Brooke (relict of Robert Brooke of Cockfield Hall, Yoxford) with an Account of her Life and Death*. London.

N Pevsner & E Radcliffe, 1975. *The Buildings of England; Suffolk*, Penguin, Harmondsworth.

C Platt, 1994. *The Great Rebuildings of Tudor and Stuart England*, UCL Press, London.

J Poster (ed.), 1986. *George Crabbe: Selected Poetry*, Carcanet, Manchester.

U Priestley, 1990. *The Fabric of Stuffs: the Norwich Textile Industry from 1565*, University of East Anglia, Norwich.

J Ravensdale, 1968. 'Landbeach in 1549: Ket's Rebellion in Miniature', in L M Munby (ed.), *East Anglian Studies*, Heffer, Cambridge.

V B Redstone, 1904. *The Ship-money Returns for the County of Suffolk 1639–40*, SRS.

V B Redstone, 1921. 'Dutch and Hugenot Settlements of Ipswich', *Proceedings of the Hugenot Society of London*, 12 (iii), 1–22.

C Richmond, 1981. *John Hopton: A Fifteenth-Century Suffolk Gentleman*, Cambridge University Press, Cambridge.

S W Rix (ed.), 1852. *Diary and Autobiography of Edmund Bohun 1677–97*, Beccles.

RCHM (Royal Commission on Historical Monuments) 1972. *County of Cambridgeshire, Vol. 2, North-East Cambridgeshire*, HMSO.

C Russell (ed.), 1973. *The Origins of the English Civil War*, Macmillan, London.

E Sandon, 1977. *Suffolk Houses*, Baron, Woodbridge.

B Schofield, 1957. 'Wreck Rolls of Leiston Abbey', in J Conway Davis (ed.), *Studies Presented to Sir Charles Hilary Jenkinson*.

B Sharp, 1980. *In Contempt of all Authority: Rural Artisans and Riot in the West of England, 1586–1660*, Berkeley & Los Angeles University.

J Sheail, 1971. *Rabbits and their history*, David & Charles, Newton Abbot.

J F D Shrewsbury, 1971. *A History of Bubonic Plague in the British Isles*, Cambridge University Press, Cambridge.

S M Sommers, 1995. 'Dunwich: The Acquisition and Maintenance of a Borough' *PSIA* 38 (iii), 317–30.

M Spufford, 1974. *Contrasting Communities: English Villagers in the Sixteenth and Seventeenth Centuries*. Cambridge University Press, Cambridge.

J Thirsk (ed.), 1967. *The Agrarian History of England and Wales, Vol. 4, 1500–1640*, Cambridge University Press, Cambridge.

N C P Tyacke, 1951. 'Migration from East Anglia to New England before 1660', Ph.D. thesis, London.

D Underdown, 1985. *Revel, Riot and Rebellion: Popular Politics and Culture*, Oxford University Press, Oxford.

F H Vertue, 1889/90. 'The Family of Gardner, the Historian of Dunwich', *East Anglian*, N.S. 3, 84–5.

J Walter, 1996. 'Anti-Popery and the Stour Valley Riots of 1642', in D Chadd (ed.), *Religious Dissent in East Anglia*, vol. III, Proceedings of the Third Symposium, Centre of East Anglian Studies, pp. 121–40.

J Walter, 1980. 'Grain riots and popular attitudes to the law: Maldon and the crisis of

Bibliography

1629', in J Brewer & J Styles (eds), *An Ungovernable People: The English and their Law in the Seventeenth and Eighteenth centuries*, Hutchinson, London.

J Walter & K Wrightson, 1976. 'Dearth and the Social Order in Early Modern England' *Past & Present*, 71, reprinted in P Slack (ed.), 1984. *Rebellion, Popular Protest and Social Order*, Cambridge University Press, Cambridge, pp. 108–28.

J Walter & J Morrill, 1985. 'Order and Disorder in the English Revolution', in A Fletcher & J Stevenson (eds), *Order and Disorder in Early Modern England*, Cambridge University Press, Cambridge, pp. 137–65.

J Walter, 1985. 'Social Responses to Dearth in Early Modern England', in J Walter & R S Schofield (eds), *Dearth and the Social Order*, CUP.

R S A Warner, 1933. *Sir Thomas Warner, pioneer of the West Indies*, West India Committee, London.

P M Warner, 1982, 'Blything Hundred: a Study in the Development of Settlement, AD 400–1400' Ph.D. thesis, Department of English Local History, University of Leicester.

C V Wedgewood, 1958. *The Great Rebellion: The King's War 1641–1647*, Collins, London.

K Wrightson 1980, 'Two concepts of order: Justices, constables and jurymen in seventeenth-century England' in J Brewer & J Styles (eds), *An Ungovernable People: The English and their Law in the Seventeenth and Eighteenth Centuries*, Hutchinson, London.

C H E White (ed.), 1883/88. 'The Journal of William Dowsing, Parliamentary Visitor Appointed to Demolish Church Ornaments, etc., within the County of Suffolk' *PSIA* 6 (ii), 236-95.

W White, 1844 (facsimile 1970). *History, Gazetteer, and Directory of Suffolk*, David & Charles, London.

A Young, 1771. *The Farmer's Kalendar* (EP Publishing, Wakefield, 1973, facsimile edition).

Note on transcriptions

Original spelling has been retained except where *the* has been substituted for *ye* and *i* for *j* in the case of numerals. In one or two instances a spelling has been changed where there is ambiguity of meaning such as *buy* for *by*, and *quay* for *key*, but only where the context makes the interpretation irrefutable. All unusual spellings which indicate dialect have been retained.

Capitalisation of letters has been modernised throughout, including the text from Gardner.

Punctuation and paragraphing have been added to aid readability, particularly in the lengthier transcripts from Gardner and the church wardens' accounts where there are no paragraphs in the original form. The paragraphing and insetting on the legal papers has been replicated as far as possible.

Surnames and place-names remain unchanged regardless of variations in spellings. Word contractions have been transcribed in full except for titles and contractions that are still in modern usage.

APPENDIX

Selected documents

Document 1: Hopton lease of 1534. HA30:50/22/3.1 [27]

This Indetrye made the xvi day of the month of July in the xxc yere of the reigne of oure Sovereign Lord King Henry the viii [1534] betwixt Syr Arthur Hopton knight of the on pertie. And Regnold Barnard gentilman Roberd Pygott marchaunt William Barrett yeman Robard Burfuard maryner And John Tompson Cordwainer of the other partie. Witnesseth that the said Syr Arthur Hopton hath Demised graunted and to ferme letyn ... to the forsed Regnold Barnard Robard Pygott etc ... All that the Sheppasture leyng betwixt Blyburgh and the town of Walberswick together with a close called Feld Clos leying in Blyburgh aforseyd. To have and to hold all the said Sheppasture and clos called New Clos to the forseid Regnold Barnard Robard Pygott etc ... from the fest of Seynt Michaell the Archaungell next to com after the date of these presents unto the End and terme of x [ten] yeres then next folowying and fully complete. And forthermore it is fully condistehdid and agreed bytween the partyes a foreseyd by the said Regnold Robard William etc ... shall have duryng all the seid terme ther eseaments of the berne within the parke of Westwood therin to leye ther corne together with the chaymbyrs ther for to leye in ther wolle and ther hors harnes. And moreover it is agreed that the seyd fermors ... shall have at all tymes sufficient of alldyres to make withall ther bridges for ther shepyes gate into the fenne. Within the Alldyrcarre of the sed Syr Arthur Hopton called Borwardes or Mekylfyldes lying in Blyburgh a forseyd. for the whitche seyd sheppasture and eseamentes of the berne chaumbyrs and alldycarr as is beforsaid theseyed Regnold Barnard Robard Pygott ect ... covenenten and graunten ... to the said Syr Authur Hopton his executores and his assignes truly to content and paye ... every year after other duryng all the seid term xx pounds of Englysh money tobe payed in maner and forme folowing that is tosey at the fest of Esterne next coming after the date herof x Li. And at the fest of Seynt Michaell the Archaungell next folowiyng x Li. And so forth every yere after other duryng theseid termes xx Li to be payed by the sayed Regnolde Barnarde Robard Pygott etc ... to the said Syr Arthur Hopton ... Provided allweyes that if in cas be it shall happen att any tyme here aft duryng the syd terme the seid Syr Arhur Hopton or his heyres to be myndyd to take in to his or ther own hondes the seid Sheppasture Berne New Close Chaumbers and Alldercarr and to leye it with his own catell or theres withoute eny fraude or dissyte. That then the seid Syr Arthur Hopton his heyre or his heyres or assignes shall geve wornyng to the forseyed Regnold Barnard Robard Pygott etc ... xii monthes next before that they or any of them shall depart from all the seyd ferme ... And the seyd Syr Arthur Hopton and his heyres shall m' to all the seyd ferm holly to reentre. this present leas nott with stondyng. In witness ...

Arthur Hopton Knt
[only the one sigature and one seal.]

Document 2: Testament of 1556. HA30:50/22/3.1 [53]

Because yt is meritorious in all thinges to tesifie a trueth. We the inhabitantes of Blytheboroughe and Walberswyck beyng assured and perfectelye knowinge a certeyne lease to be made from Syr Arthure Hopton Knyghte and Owen Hoopton Esquier to Syr Edmund Rous Knyght in the xxxijj yere of the reygne of our late Soveraine Lord

(1540–1)   Kynge Henrye the eyght, for the foolde course and sheipe walke byinge btwene Blythboroughe and Walberswyck withe all lands and marshes to the same belongynge do testifye and assyrterne that the sayed Syr Edmund Rousd did peaceablye and quyetly enioye posses manure and feide the sayed walk and sheipes pasture; by the space of fyve yeres and more next ensuynge the date above wrytten; And certeyn of the same grounds put in tilthe and beinge sowen with barley; the sayed Syr Edmund Rous by hys servints dyd take our cattell fedynge towarde the same corne. and dyd empounde them; for the damages wherof, suche unreasonable charges were answered contentyd and payed to the sayed Syr Edmund Rous hys servants as they would requyre & & Not withstandynge that we the sayed Inhabitantes of Walberswick had a former lease granted to us by the sayed Syr Arthur Hoopton Knyght in the xxvth yere of the Raygne of the sayed kynges

(1533–4)   maiestie: yet beying afterward required by the sayed Syr Arthure Hoopton Knyght And Owen Hoopton Esquire, to yelde upe our ryght and tytle, to the sayed Syr Edmund Rous and his assignyes we the sayed inhabitates havynge ij hoole yeres and more in our sayed lease dyd yelde upe the sayed lease with all and singler the premisses to our great losse and hyndtrans for we would not suffer the sayed Syr Arthure Hoopton Knyght And Owen Hoopton Es-quire to forfeite theyr londe made to syr Edmund Rouse; as allso we afferme and testifye that the sayed Owen Hoopton Esquire in the xxxvijth yere of the Raygne of the

(1540–1)   sayed Kynge his Mtie. dyd bye of the sayed Syr Edmund Rous the whole number of [sheep] As they were fedynge in and upon the sayed walke together with the lease to hyme made by the sayed Syr Arthur Hoopton Knyght and Owen Hoopeton esquire as ys afore sayed –

In wytness wherof we the sayed inhabitants of Blythborough and Walberswyck have sette oure handes and sceales dated the xj of

(1556)   March In the scecond and thyrd yere of the Raygne of Phylip and Marye by the grace of god kynge and quene of Englonde France

Naples Jherusalem and Yrelond; defenders of the fayth Prynce of
Spayne and Cevile Archdukes of Austrich Duke of Myllayne
Burgundie and Braband, Countes of Haspurge Flanders and Tyroll
& & &

me Wylliam fekett

… Hillston	Robert Renersham	Thomas West	Thomas Smyth
Alexander Richsone	Edmund Hayloke	John Freer	Thom …
Jefrey	Thomas Chamburs	Lawrence Drane	…
by me Roberd myllys	Robert hdthu …	Nythanl crowe	John R …

Document 3: HA30:50/22/3.1 [50]

Walberswicke   The 19th of december 1636
in suff.   The manner of the rape that William Turrould committed
with Margarett the wife of Tho: Sherringe: and Alice the
wife of William Sallows as it was related to me Robt. Tanner
by Robt. Edmunds and his wife and the said Margarett
Sherringe and Alice Sallowes all together:–

That the said William Turrould in an eveninge: in the end of the month of
November 1635: did com to the house of one Robt. Edmunds of the town
aforesaid the dore being shutt: and did drawe the latch and open the doore
and went in and shett the dore againe there beinge in the house at the present
Margarett the wife of Tho: Sherringe and Alice the wife of William Sallows
daughter to the said Robt. Edmunds and no bodie else save a infant of the
said Sallowes: and then and there did with his yard drawne violentlie put his
hand under the clothes of the said Margarett Sherringe and did inforce hir
to have comitted ffilthines with hir: who striving with him to hir power yett
could not ffre hirself from him called to the other woman for help: She layinge
hir infant which she had in hir armes uppon the harth did with sped runne
to help the said Margarett Sherringe: and laiyinge hold of the said Turrould:
the said Turrould leaving the said Margarett did laye hold of the said Alice
Sallows and gatt hir downe as he had done the other woman [and] did strive
also with hir in so much that she was constrained to call out for help: the
other woman beinge out of breth the mother of the said Alice Sallowes being
within hearringe cam runninge into the house would have pulled the said
Turrold from hir daughter but could not: insomuch that she was inforced to
take a staffe and did strike the said Turrould therewith before she could make
him forbeare this abominable acte.

and as convenientlie as they could did call him the said Turrould before
John Scrivener Esq. one of his majesties Justices of the peace who did binde
the said William Turrould to the next quarter Sessions upon the information
that the said goodwife Edmunds, Alice Sallowes and Margarett Sherringe did

give unto the said Mr Scrivener uppon their oathes: which was rather more vile than is above spasefied but when these three women cam at the sessions to give evidence against the said Turrould: Mr Henry Cook one of his majesties justices being one of the bench at those sessions would not suffer the wittnesses to give in there evidence and said there ware none but roagues and whores and theves that cam against him in so much that there was nothinge don against the said Turrould and hereunto I give my oath if I be there unto called.

me Robt. Tanner

the 2 younge women are both liveinge still.

Document 4: William Turrould's petition of 1637. HA30:50/22/3.1 [1]

To the Kings most Excellent Majestie

In all humble manner doe complaine to your majestie. William Turrold & sundry others your highnes loyall poore subjects the tenents of the manor of Walberswick in your majesties County of Suffolk to the number of 300 concerning the greate wrongs done unto them by Sr Robert Brook Knt Lord of the said manor as well in exacting excessive ffynes upon admission of Tennants being 10 times as much as have been used to be paid, as alsoe in debarring them from such common of pasture for their cattle as they have time out of mind held & enjoyed, and by unjust detayneing from them certain marsh grounds which of right belongeth unto them, upon which common & marsh grounds the said Sr Robert Brook hath caused a great ffarme house to be built & hath enclosed in a great pert of the said marsh grounds & common & laid it to the said ffarme house & therein placed tennants of whom he receiveth much rent to the disinherison of your majesties poore subjects, And the said towne being a haven towne on the seaside he the said Sr Robert Brook hath taken away from your petitioners an ancient key [quay] belonging to the towne the benefit whereof in former times was imployed for the repaire of the church & releife of the poore, but now the said Sr Robert taketh all the benefitt therof to himselfe, soe that your poor petitioners by the incloseing of the common & marsh grounds & by the losse of the said key are wronged & dammised £200 per annum. And the said Sr Robert Brook doth with beasts, sheepe & conies so overcharge the rest of the common not inclosed that they breake into the petitioners inclosed grounds & eat up & destroy their grasse & corne & what they sow & plant they cannot reape & enjoy. By which cruell & unjust dealings of the said Sr Robert your majesties poore petitioners have for many years groaned under the burden therof, & many of them have beene constrayned at a great undervalue to sell away their land and to leave the country.

Now forasmuch as your majesties petitioners by reason of their extreame povertie & perversnes of the said Sr Robert are of themselves unable to redresse these intollerable wronges by the cause of your majesties lawes

May it therfore please your majestie to refer to the premises unto Sr Thomas Glemham, Sr William Springe, Sr Roger North, Sr Edmund Bacon, Sr Robert Coke & Sr Butts Bacon Knights & to Henry North, Henry Coke & John Scrivener Esquires or to any 4 or more of them who are all acquainted with these aforesaid oppressions & that they may have full power to call before them, & examine any persons whatsoever for the discoverie of the premises & make a final end therof if they can & restore every man his right, or else certifie your majestie the state of the cause that such further cause may be taken for your petitioners releife as in justisshalbe thought fitt.

And your poore subjects (as in duty bound) will dayly pray for your majesties longe & prosperous raigne ever as.

At the Court at Grenwich 23rd Jani 1637

His majestie is gratiously pleased to referr this peticon to the committies desired or to any 4 or more of them to the end; they may acquaint the said Sr Robert Brook therewith & require his presence & meeting with them at such dayes & places as they shall appoint and to inform themselves from the said Sr Robert of his answere to the matters complayned of, and also to mediate with the said Sr Robert soe compasing the differences between him and the petitioners; and for releife of the petitioners if they finde just cause, or otherwise to certifie his majestie their proceedings & the true state of this business & their opinyons touching the same.

Granted by Sr. Sidney Mountague.

Document 5: Sir Robert Brooke's answer (Part a). HA30:50/22/3.1 [42]

That the complainant is owner of the mannor of Blythburgh cum Wallberswick, & that Paules Fenne & East Marsh, and 500 acres of land called the Sheepswalke abutting the heath are within that mannor.

Exparle quer'

That the plainant is owner of a marsh in Walberswick called East Marsh conteyning 16 acres or therabouts, & that the same hath byn reputed & taken as parcell of the said mannor, and that the plainant & those whose estate he hath his & their farmers have quietly injoyed the same, & taken the profitte thereof.

Jo Skinner fol 28 Int 2 & 3 for 60 years
Sym Austen fol 1 Int 2 7 3 for 50 years
Thomas Johnson fol 9 Int 3 for 50 years
Alice Tidsdale fol 15 Int 2 & 3 for 50 years

That the plainant is owner of a peice of marsh called Paules Fenn being 26 acres, & that the fenne hath byn reputed & taken as parcell of the said mannor, and that the plainant his ancesores, his & their farmors of Westood Lodge have quietly injoyed the same and taken the profitts therof. And have know divers farmors that have occuied the same.

> William Stannard fol 22 Int 2 & 3 for 68 years
> Ro. Skynner fol 28 Int 2 & 3 for 60 years
> Sym Austen fol 1 Int 2 7 3 for 50 years
> Thomas Johnson fol 9 Int 3 for 50 years
> Alice Tidsdale fol 15 Int 2 & 3 for 50 years

That when any of Walberswick town cattle have come into the Paules Fen & taken feeding there, they have been driven out & impounded.

> William Stannard fol 22 Int 3

That the Paules fen lyeth in Blyburgh. [Six tenants listed with names as above]

That Paules fen have been rated to Blyburgh.

> Sym: Austen
> Arthur Styles

That the said Paules Fenn & East Marsh were inclosed by a wall from an arme of the sea about 40 or 50 years since by the plainant or his father, and before that time they were not worth above xijd. the acre per annum. That the incloseing of them cost one hundred or two of pounds. [Four tenants listed with names as above]

The Heath

That the plainant & those whose estate he hath, his & their farmors have had two foldcourses & a warren of conyes within the said mannor of Blythburgh cum Walberswick, upon the sheepwalk or heath there conteyning 500 acres, and that the said conyes & sheep have fed all over the said heath unto the very towne of Walberswick.

> [Eight tenants listed as above with Jo Clarke and Isaack Johnson]

That the lordes of the said manors or their farmers have used to plow such part of the said (walk or) heath as they would, & when any part therof was sowen with corne the inhabitants of Walberswick did not put their cattle upon any such places soe sowen untill the corne was reaped, but if their cattle did stray & come on the corne they were impounded. And that it appeares by the rigges & furrowes on most parte of the heath, that the same have usually byn plowed.

> [Six tenants listed with names as above]

That the inhabitants used to have a follower with their great beasts when they put them upon the said walke or heath, to keep them from straying.

Sir Robert Brooke's answer (Part b) HA30:50/22/3.1 [51]

The Answere of Sr Robert Brooke knight to the unjust complaints of William Turrold & others contayned in a Petition by them exhibited to the Kings most excellent Majestie At the Court at Grenwich the 23 of June 1637

Imprimis this respondent saith that he is seised of the manor of Blythburgh within the county of suffolk being holden of his Majesty by knights service in capite unto which Walberswick & Hinton are hamblettes and members and the mesuages landes & tenements within the same are holden of him as of his said Manor of Blythburgh and that there is not an hundred tenents to them all.

Excessive fines	Concerneinge the exacting of excessive fflynes upon the admission of tennants fines which they pretent to be ten tymes soe much as have been usually paid he saith that the fynes are arbitrable at the will of the Lord yet he hath not taken two yeares proffitte of any one tenament for a ffyne according to the ful valew.
Surcharging the common	As touching the surchargeing of such grounde as they pretend to be their common and to eate the up what they they sow and plant. This Respondent saith that he conceiveth the same to be upon his charter warren & sheepcourse part & belonging to his manor in which albeit he doth not deny them feed for their great cattle with a follower when it is sowen with any kynd of grayne, yet he & all those whose estate he hath therin and in the said manor might lawfully at his and their will & pleasure either feed the same with his flock of sheepe replennish & keep the same for warren or plough up the same at his (pleasure) And when the same or such parte therof is sowen with any kynd of grayne that then & soe long they ought not to have any feed for their great Beasts in any places soe sown.
for the taking away of marshe grounde	And as touching the taking away of certain Marsh grounds wich they pretend to be their common this respondent saith the same hath allways been used & fed to Westwood Parke & occupied therewith & is parte of the demesnes of of his said manor of Blythbrough & lye within the bounds & lymitts of the towne of Blythbrough. And was inclosed (above 40 tie years now last past) with a bank from the sea by his father, and he further saith that since his said fathers death purchased of one Mr Cannon one other marsh con-

tayneing 16 acres (or there abouts which lyeth next unto the marsh called Paules Fen the same 16 acres doth) abbutt (as appeareth by the old tinderse) upon the marsh of the lord of the said mannor which is called Paules Fenne.

for the farme house pretended to be bilt upon the common

Concerning the Farmhouse wich they pretend is built upon their common the same is built upon this respondents severall freehold wich he purchased.

For the Key [quay]

For the key wich they alleadge is taken from them that cause amoungst others now depending before the barrons of his majesties Court of Exchecquer where this respondent hopeth it wil be determyned.

Whereas they complaine that diverse of them by his respondents unjust dealings with them have been constrayned at great undervalue to sell away their land & to leave the country. He utterly denyeth that he hath byn the cause that any man should soe doe.

He humbly desireth the benefit of a subject for preservacon of his inheritance according to law.

Document 6: The tenants' reply. HA30:50/22/3.1 [43]

It's granted that the complainant is owner of the manor of Blythburgh cum Walberswick, and that the defendents are tenants.

Ex part defendants:
That the tennants of the said mannor inhabiting within the town of Walberswick have used to feed their great cattle of all sorts at all times in the yeare in a marsh called Pauls Fenne conteyning 60 acres lying in Walberswick and in like manner have fed their cattle in a marsh called East Marsh conteyninge 20 acres lying in Walberswick.

> Robert Harte fol 53 Int 1 & 7 for 63 years
> Christopher Frost fol 63 Int 1 & 7 for 63 years
> Ro Durrant fol 1 Int 1 & 7 for 60 years
> William Blyth fol 7 Int 1 & 7 for 60 years
> Dorathy Crones fol 59 Int 1 & 7 for 53 years
> Anth Gester fol 70 Int 1 & 7 for 50 years
> Ro Freeman fol 16 Int 1 & 7 for 50 years
> Nath Tredescant fol 41 Int 1 & 7 for 46 years
> Marg Peterson fol 34 Int 1 & 7 for 42 years
> Edw Crow fol 22 Int 1 & 7 for 40 years
> James Stacy fol 27 Int 1 & 7 for 40 years

That the said Pauls Fenn lye in Walberswick & have from time to time made and mayntayned causyes, hanges & passages for their cattle to goe & passe into the said East Marsh and Paules Fenne.

[Eight tenants listed with same names as above]

That Pauls Fen & East Marsh were inclosed with a wall by the complainants agents, but the complainants father did not inclose the same when he walled in other marshes, which might then have byn done with lesse charge.

[Three tenants listed with same names as above]

That they have a town booke wherein the charges of makeing passages & draynes unto & in the said Paules Fen are set downe to be done at the cost of the inhabitants.

John Barwick fol 50 Int 8
Edw Burford fol 51 Int 8

All the Wittnesses depose that the tennants & Inhabitants of Walberswick for themselves & under tennants for all their great beasts wich they can keep on their owne lands used to have common of pasture upon the great heath or common called Walberswick Common at all tymes in the yeare, and that the said heath extendeth from the townes end of Walberswick to Deadmans Crosse. That the said tennats & Inhabitants have usually fed their sheep with a shepheard upon the said heath.

[Ten tenants listed with names as above]

The number of their sheep have been 3 or 4 hundred.

[Three tenants listed]

When any of the said heath have been plowed & sowen with corne by the lord or lords of the said manor or their under tennants that great part of the said corn have been reaped & carryed away by the commoners of Walberswick.

[Eight tenants listed with names as above]

That divers of the inhabitants did frequently in the day time with doggs & netts take connyes wich did burrow & breed on the heath. [Five tenants listed as above] They all depose that the tennants of the manor of Blythburgh cum Walberswick have several wastes wherein the tennants of Blythburgh have common by themselves, & the tennants of Walberswick by themselves.

Document 7: Articles of Misdemeaner against William Turrold.
HA30:50/22/3.1 [52]

Articles of Misdemeaner & Liberties againstt Willm Turrold of Walberswick in the countie of Suffolk, Miller.

Imprimis: the said William Turrold came walking fom his owne house unto the mill of Walberswick & went into the mill & there fetched out a quarter staff having grynes on the end & a pike in the other end & a sword by his side & came from theare unto Thomas Mayhew & George-Walter-Cricker who were working uppon a baulk nere the said mill & said unto them I foresend you walkag any longer here if you doe I will kill you if there were no more men in England & forthwith made at the said Thomas Mayhew with a thrust & after came on & srooke him five or six blows with the same staff with which blowes he brake the said staffe & when he had done so he drew his sword at the said George Cricker & swore by the Light of god that if the ~~said~~ they oray other man new of it were therr great Mr Sr Robert Brocke came there eithr by night or day he would shoote a bullett in his side if he were hanged even an hour after.

Item: the said Wm Tirrold hath lately abused (Margaret) the wife of Thomas Sherringe (&) did endevour to have eabusehed (her) [abused her] ~~had not he been prevented by company that came in~~ & used great violence upon her & threw her downe to the ground & would have done his pleasure upon her had he not bene prevented by company that was called in to help her

Item: the said Wm Tirrold ~~didch dayly~~ did lately walk up and downe the towne of Walberswick ~~in the daytime~~ with his sword by his side & will not suffer any execution or process h ... ilrie upon him to the great terror of the poore inhabitants in Walberswick.

...	=	text missing
( )	=	text interlined or inserted

Document 8: William Turrold's second petition: May 1638.
HA30:50/22/3.1 [41]

To the Kings most excellent majestie

The humble peticion of William Turrold & diverse other your majesties poor subjects & inhabitants of Walberswicke County of Suffolk.

Humblie sheweth that your petitioners haveing hertofore exhibited their humble petition unto your Royall Majestie complayneing of many wrongs done unto them by Sr Robert Brooke Knt lord of the manor of Walberswick. And your majestie was graciously pleased to refere the examinacion of your peticioners said complaint unto Sr Thomas Glemham Sr William Springe and the rest of the referres mencioned in the same peticion hereunto annexed to the end they or any fower of them might mediate with the said Sr Robert

Brooke for composeing the differences & for releife of your peticioners or
otherwise to certifie your majesie of their proceedings and the true state of
the busines together with their opnyons touching the same

So it please your majestie the said comittees have taken an exact examinacion of the whole estate of the busines, they have also mediated with
the said Sr Robert Brooke for a peaceable end, to which the committees
have found your peticioners very inclyneable yet in respect of the said Sr
Robert there labour & indeavors hath been but in vayne wherefore in
obedience to your majesties royall derecions they have certified the true
state of the busines under their hand hereunto annexed

The premisses considered in regard the poore inhabitants of the said
towne of Walberswick hath been greatly impoverished by the op-
pression of the said Sr Robert Brooke, yet your peticioners humble
suite is that your majestie will be graciouslie pleased to referre the
re-eexaminacion of the busines unto the Lord Arch Bishopp of Canterbury his grace the right honourable the Lord Keeper & the Lord
Bishop of Norwich authorizing them or any two of them to call the
said Sr Robert Brooke before them and to make sure a fynall &
determinate end of the busines for your peticioners releife as shall be
agreable to equitie and right without any further trouble to your
majestie herein

And your poore subjects (as in duty bound) will daylie pray for your Majesties
Long & prosperous raigne over us.

Att the Court at Whitehall 2 Maii 1638

His Majestie is pleased to referere the consideration of this peticion and the
annexed to the Lord Arch Bishopp of Canterbury his grace & the Lord Keeper
to the intent their lordshipps upon examinacon of the premisses may settle
such cause therin as shalbe agreeable to equitie & good conscience.

Edw: Powell

We appoynt Wednesday the sxt of June next for the hearing of this
businesse at the counsell board and doe hereby will & require Sr Robert
Brooke within menconed or any else whome it may concerne by
themselves in person or some others sufficiently instructed and auth-
orized by them to attend accordingly. Provided that tymely notice be
given & a true coppie of this peticion & referrence delivered to them./

8 Maii   Wm Cant:
1638   Tho: Coventry.

Document 9: The petition delayed: June 1638. HA30:50/22/3.1 [40]

To the most reverend father in God William Lord Arch Bishop of Canterbury his grace & the right honourable Thomas Lord Covetry Lord Keeper of the great Seale of England &c.

> The humble petition of William Turrold & diverse other the poore Inhabitants of Walberswick in the County of Suffolk

Humbly sheweth

That the petition of your Lorshps petitioners presented to his majestie for diverse and sundry wrongs by them susteyned by the iniust dealings of Sr Robert Brooke knight lord of the manor of Walberswick aforesaid his majestie was graciously pleased to referre the consideration of the greviances and abuses conteyned in the said peticion to your Graces 7 Lordships for the examinacon of the premisses & the setling of such cause therein as should be agreeable to equity & good conscience as by the same petition annexsed may appeare.

And whereas your Grace and Lordship was pleased to appoynt Wednesday the 6. of this instant June for the heareing of the said Cause

> Soe it is may please your grace & lordship that your said petitioner Turrold & others of the said inhabitants attended both that Wednesday & two dayes after with their cousell to their great charges & expences (the same being a poore marytane towne) but could not be heard.

> Wherefore your petitioners most humbly crave your grace & lord-ship's favour that you wil be pleas'd to assigne & appoint some other certaine time in the beginninge of the next Michaelmas terme for the examinacion of the premisses to the end that your petitioners & the said Sr Robert Brooke may againe give theire attendance upon your lordhsips the rather for that theire aboad is farre distant from this place, and as in duty bound they will ever pray &c.

We appoynt wednesday the seventeenth of October next for the hearing of this busines at the counsaile table in the afternoone & doe hereby will and require that upon notice given, all partyes whom it concernes by themselves or their Counsell be ready to attend accordingly.

> June 16   W: Cant: Tho: Coventry:
> 1638

Document 10a: Star Chamber proceedings, 3 May 1639. HA30: 50/22/3.1 [41]

To the most reverend father in god William Lord Archbishop of Canterbury his grace, & the right Honorable Thomas Lord Coventrye Lord Keeper of the grete seale of England.

The humble petition of William Turrold & diverse others the poore
Inhabitants of Walberswicke in the County of Suff:

Humbly sheweth

That uppon the humble petition of your Lordships petitioners presented to
his sacred majeestie for divers sundrie wrongs by them systeyned by the uniust
dealings of Sr Rbte Brooke Kt, lord of the manner of Walberswick aforesaid
his majesie was gratiouslie pleased to referr the consideration of the grevances,
abuses conteyned in the said petition, to your grace & Lordship: for the
examination of the premisses & the setting of such course therein as should
be agreeable to equitie & good concideration.

And whereas your grace & lordship was pleased to appoint severall dayes
for the hearing the said cause, as by the said petitioners here unto annexed
may appere, att which several tymes your said petitioner Turrold and
others of the said Inhabitants attended with their cause to the great charges
& expences, the same being a poore maritane towne butt could not be
heard, whereby your poore petitioners are much disinabled.

May it therefore please youre grace & lordshipps favour to assigne &
appoint so certaine tyme in the beginning of the differences to the
end that your petitioners and the said Sr Robt Brooke may againe give
there attendence uppon your lordships and your petitioners shall humbly pray for your lordships helth & happiness long to continue.

In regard the former days appointed did not hold, we appoint Fryday the
third of May next for the hearing of this business, & do hereby require that
notice be given to all parties concerned, & that by themselves or there counsell
thay attend accordingly.

W: Cant   Tho: Coventrye.

[on the reverse]

At ye Inner Starchamber 3 May 1639
Their Lordshipps have appointed Friday next being the tenth of this Month
to heare this Businesse about fower a clock in ye afternoone ye Inner
Starchamber, and require that all parties concerned herein have timely notice
given them and that they attend accordingly themselves or their councells

Edw. Nicholas [clerk]

Document 10b: HA30: 50/22/3.1 [46]

At ye Inner Starchamber 22 May 1639

Present.

Lo: Arch. Bop. of Cant, Lord Keeper

Their lordships having this day heard the councell as well of the inhabitants of Walberswick i ye county of Suffolk as of Sr Robert Brooke concerning several differences between them. It was by their lordships ordered, that as concerning the right of commonage which the said town clayme to have in certaine upgrounds and fens in and neare ye said Town being a matter of right and determinable by lawe, that the said Inhabitants shall cause an action to be brought against ye said Sr Robert Brooke, to which the said Sr Robert shall appeare forthwith that a triall may be had at ye next assize which triall their lordshipps order shalbe for that matter finall. As concerning ye key [quay] in Walberswick it is agreed on all sides that the right thereof belongeth to ye said towne, but whereas ye said Sr Robert hath been at charge of ye reparacon therof, it is by their lordships ordered that a commission shalbee forthwith issued under ye great seale directed to Sr Butts Bacon Bart: Sr John Rous Knt, Henry North and John Scrivener Esq. authorising them or any two or more of them, by examinacon of witnesses upon oath or otherwise, to find out what & how much Sir Ro. Brooke hath disbursed in repayringye said key, and what rent hath been by him receaved for ye same and how much of ye money by him disbursed so thereby reimbursed, and to make certificate thereof to their lordships to which all parties present did consent, and if ye said commissioners cannot otherwise make an agreement between ye said townsmen & Sr Robert Brook, then their lordships upon receipt of ye said certificate will give such further order as shalbee just and fit for determining ye said diffferences.

Ex Edw. Nicholas [clerk]

Document 11: Sir Robert Brooke petitions for more time: June 1639. HA30: 50/22/3.12 [39]

To his most Reverend father in God William Lord Archbishop of Canterbury his grace Primate of all England &c. and to the right honourable Thomas Lord Coventry Lord Keeper of the Great Seal of England

The humble petition of Sr Robert Brooke Knight.

shewinge

That whereas upon a reference from his Majestie there was an order made at the Inner Starrchamber 22 Maii last that the inhabitants of Walberswick in Suffolk between whome & your petitioner there is a diuference for right of commonage would cause an action to be brought against your petitioner to which your petitioner should appeare forthwith that so a tryall migh be had at the next assized your petitioner is most willing to obey the sayd order: but in regard he was not served with nor saw the sayd order untill the 24th of this instant June At which tyme he was served with the same at his house in Suffolk neare 80 myles from London his scollicitor being then in London and for that the tyme is now so short & your petitioner so unprepared that he cannot be ready for a tryall at this assized.

He humbly beseacheth your grace & your good lorship to appoint & Bloody Marsh assigne some further tyme for the sayd tryall at the next assizes in Lent.

And he shall pray &c.

Document 12: Brooke's appeal & order for a re-trial: 19 August 1640. HA:50/22/3.1 [26 & 38]

Att Whitehall the 19th: of August 1640

present

Lo: Arch Bp: of Cant:	Ea: of Holland
Lo: Keeper	Lo: Lieut of Ireland
Lo: Trearer	Lo: Cottington
Ea: Marshall	Mr: Sacr Windebank
Lo: Chamblanne	Sr: Tho: Rowe
Ea: of Dorsett	Lo: ch: Justice of the Common Pleas

Whereas an humble Petition was this day read at the Board in the name of Sr Robert Brooke Kt, shewing that by an order made at the counsell board the 22nd of May 1639 – here in Michaelmas Tearme last appeared to an accord brought by one Edward Howlett for the tryall of the right of commonage clymed by divers of the inhabitants of the towne of Walberswick in the County of Suffolk, in three severall peices of the petitioners land, at which tryall (by the same order) a finall settlement touching that matter was appointed: further shewing that at the last sommer assizes for the said county of Suffolk: the said cause was tryed before Sr Edward Littleton Kt, Lord chief Justice of the Common Pleas a member of the board, which cause being popular the jury (being divers of them returned upon tales) gave a verdict against the petitioner, contrary to the expectation of those that heard the evidence, which if the petitioner be concluded therby, will prove very preiudiciall unto him, hee humbly praying their lorships to heare the opinion of the said Lord chiefe Justice therein, and to be releeved by a new tryall. – Their Lordships upon consideration had thereof, as also having heared the said Lord Chief Justice now present in counsell, expressing, that according as the busines lay before him at the said assize, he could not be of other opinion, but that it was requisite & fitting, the petitioner should have another tryall: It was therefore ordered according to the humble suite and desire of the petitioner, that he should not be barred from having such new tryall as aforesaid without any preiudice by reason of the aforesaid order of the 22nd of May 1639 – and the said Lord chief Justice is hereby prayed to make such stay of the postes & all proceedings upon the verdict given at the last assizes for the county of Suffolk against the said Sr Robert Brooke touching this mater as his Lordship shall think fitting.

Ex, D: Carleton.

Document 13: John Barwick's story of 1652 abstracted by Thomas Gardner in 1754 from the churchwardens' account book (FC185/E1/2). (Gardner 1754, pp. 123–6)

1642. In a trial before Judge Littleton, the inhabitants of Walberswick regained the quay, detained ten years; and their commons containing above one thousand four hundred acres, together with the fenns, withheld from them upwards of thirty years, by Sir Robert Brook, lord of the manor; but the peaceful enjoyment thereof lasted not full two years, great troubles arising by means of Thomas Palmer, Sir Robert's great farmer, feeding not only the uplands with his sheep, but by putting them also into East-Marsh, which the townsmen having fenced with posts and rails, and scoured the dikes, the first were cut down and broke to pieces, and the latter filled up with the manure thrown out before, supposed to be done, chiefly, by means of the said Palmer.

Also, Sir Robert set up a boarded house for men and dogs, near Pauls-Fenn, to keep out and drive away any cattle belonging to the town of Walberswick; when one of the keepers came into the said town, and quarrelling with the townsmen, a lamentable fray ensued in which four men lost their lives, which gave occasion for calling the fenn afterwards, BLOODY MARSH.

And thus the commons and marshes were repossessed by Sir Robert as long as he lived; and being succeeded by his son John Brook Esq., he held the same and had a trial (about twelve years from[after] that before Judge Littleton) with the town, when Judge Fessant gave it in favour of the lord of the manor, who held possession thereof during his life, which was but short.

Under the virtuous Lady Brook they again enjoyed their ancient privilages, until dispossessed by one of her successors.

After the death of old Sir Robert Brooke, Knight, and lord of the manor of Walberswick, his eldest son, master John Brooke, Esq., suckceded his rom for a fewe yeares; for he dyed a veary younge man, and yet a very great troubler, an oppressor of his poore tenauntes in Walberswicke, in drivinge of their poore cattle foe furre to the pound, from off their common, and there sooe longe often times continewed, that after gatt out through great means usinge, and lardge paymentes, yet the poore owners lost them soon after gatt out, both horse and cowe beastes, bygetting such a dourge in the beastly myresen place of Blyberowe pound, that soone after dyed.

Thus envious was he to his poore towne; and much more otherwayes, as in troubling and lawinge of poore men; especially those that were forward in lookinge for the towne rights from his father, Sir Robert Brooke, as many Johnes there were that did sooe, as John Fayerclife, John Arnold, John Chapman, John Barwick, John Chattle, John Aims, John Marshall, John Chappett, John Prettey, John Flick, John Rowince, and others, weare willinge for there towne rightes, which gladly would have had as in time past every poore famyly had by their common, tell after, first by Sir Robert Brooke, was taken away, and then the poore towne much decayed, and was not able to subsist; but was releaved by neere twenty townes in the country; tell at last, after petitioninge, we had a tryall with Sir Robert, and recovered our commons as once before.

And after we had possession in Paules Fenn neer two yeares, with the uplands, and began to fynd some revivinges, Sir Robert seat up a boarded howse neere Paules-Fenn, wherein he kept in men and great dogges, day and night, to dispossess the poor inhabitauntes of there right of commonadge, and soe did by violence take it awaye, and had it from us to his death; and after him sooe lykwise did his sonn, Mr John Brooke, imitate him upon the upland commons that weare plowed, in not suffering our poore gleeners to have quiet gleaninge.

But ever since Mr. John Brooke came to the livinge, he sought opportunitie to geet a tryall passe against us; and at last obteyned one throwe that sowle way they went, which neyther he nor his father coud geet beefore. But when Judge Feassant red the surquit, whoe was nerre kyndsman to the Lady Brooke, then through such fowle meanes as was used, gatt the Day against John Chapman, whoe the esquire sud for eatinge up of his, 800 acares of heath ground; this was his declaration; though he silldom had any cattell there, though the common heath grounds consisteth of more than 1400 Acres.

So that now overthrowinge this John Chapman, by such fowle means, as makinge soe many friends, a foreman of the jury was Mr William Buckenhames neere kyndiman, which William was the esquire's Steward of his coarts, and allso one Bartholomewe Bullen, whose father had bin Sir Robert Brookes bally a long while, and when he dyed, he left a great estat to his eldist Sonne Bartholomewe, who married the sayd William's Sister, and swaggered it soone away, and becam veary poore in Yoxfird, where he lived amonest his good friends, that much helped him and his; this man at this tyme was one of his 12 witnesses, whoe tooke great pains in rydinge fure and neere to make up his sume, in procuringe such lyke unto himselfe, which was a pitteous one. And allsoe a woman, who had twelve yeares before witnessed for the towne, and was one of the wittnesses that than cam first for the towne before Judge Littleton, which judge having heard six of them, would heare no more; said that six in enow, soe that our other able wittnesses weare not heard.

This woman, I say, upon some relation, she having here eldist son livinge and using a farme of the esquires in Walberswicke, was drawn to be a wittnesse for the Esquire against the poore Towne, and woman was one of the greatest Wittnesses that John Chapman had; and allsoe one old steward of Yoxford, such an other wittness as this woman whoe a little before his death spoke straung things, saying, that he spoke as Robert Turner bad him: And when the woman at the Assises had sworn against the Towne, for the esquire, then presently our lawyeare Mr. William Emingham stood up, and showed the Judge Fessant what oath she had taken for the towne 12 yeares before; then when the judge had seene it, the judge called for the woman; but the woman was goone, and proved sick that she could not com.

And other such like wittnesses neer and furr off, they had that same this Bartholomew fitched from neer London, which made up that number of 12 wittnesses, such as they weare; and our poore towne had but six, in regard that 12 yeares before, when we had then 12 able Wittnesses, that Judge Littleton would heare but six of them, sayinge, that six was enowe after he had seen the towne aunchant bookes, which did playnly showe what cost the church-wardens had bine out in those dayes about the common, as in making the way into Paules Fenn, and allsoe in dickinge and drayninge of Paules Fenn: This Judge Littleton would heare no more, and for this cause, and to spare chardge, this John Chapman had but six wittnesses there then; and though they weare more worth than six score of the 12 that cam against them.

For they would speake of the previlidge the towne had in Sir Owen Hopton's dayes, who was lord of the mannor before the Brookes cam to it, as old Robert Dorant, of Coatby, and old Robert Hart, of Hynsted, and old Widowe Wynes, of Soul [Southwould], and others born in Walberswicke, spoake boldly to the Judge Fessant and tould him, that in Sir Owen Hopton's Dayes they had nooe such dooinge for he was a worthie jentleman, and loved the poore towne, and joyed the previlidge they had by there common; and Robert Dourant, who was the townes neattards boy in Sir Owen Hopton's Dayes, have often sayd, that Sir Owen himselfe have com downe from his Westwood Lodge to Paules-Fenn and had him com up to the lodge and drynke, soe that it was otherwise in those days, than have been since.

These six witnesses were soe forward in speaking, as this Judge Pheasant would not suffer them to spake up that which they had to spake; but when the Esquirs wittnesses spoake and when they could not well speak, he gave sufferance for them to be helped by their councell; and when that jury went forth, that Judge Fessant sayd to them: 'jury take notice of the witnesses, that there are 12 against 6'. Soe they went on, and found it for the esquir; then when they had soe doone, many of the jurymen seeing a man which they had thought had bine he that had promised them soe much a piece when they had done, but when they see they weare mistaken, they slanke away; and that man, he revilinge them, tellinge them they weare rogues, in wronginge of a poore towne, and nowe would yow be rewarded sayd: 'the gallowes is their just reward'. Thus the poore towne had the overthrowe, in [the] overthrowinge [of] John Chapman.

Then next after, the esquir areasted John Barwick [the writer], declaringe, that he had eaten up with his great beasts, his hundred acres of heath ground, and would have had his tryall the winter followinge; but it being too tedious a season, and hard journey, soe furre and so fowle for such aged lame people, obteyned to have a mylder tyme for them, as at mydsomer next after; and for this cause, in my stead, John Chapman our brewer rod up with all speed to London, on my brother Roger Barwicke's mare, beinge then with us, and no such to perform it could be had, for he was to goe with speed, and soe did; though to no powerpose, for when we thought we should have goone at that mydsomer upon tryall, John Barwicke was prevented by demoure, soe nothinge done.

And yet his divilish envy began agayn with newe troubles, for upon the

10th of November 1651, the esquir arested John Barwick by the bally Dowtey, for gooinge our aunchant Bounds; Mr Benjamin Bonnes, Steward of Soul [Southwold], my authorney herein. The 18th of May 1652, by John Barwicke his order and appointment, a writ was served on the esquir John Brooke, for the commons by Mr. Tuttle, the shrive [sheriff] himself; William Gymingham my attorney, to be tryed this mydsomer, 1652. But was agayne prevented by demoure. The 28th of May, 1652, I John Barwicke was [again] arested by the bally Brytton, of Coatby, by the Appointment of John Brooke, Esquir; the Bally knowe not wherfore. Mr. Bonnes my Atorney.

Thus this yonge man, and great troubler, ply'd it for his tyme, in bringeinge of his wicked actions and endes to passe; as when he, through that fowle way, gatt the tryall passe one his syd for our commons, that made him then insult over us duringe that small tyme he had after, as in exstreemely opressinge and tearyfyinge of his pure tenaunts; and foe was like to have bine, but that God prevented him by sudden death; which death cam sooner than his expectation was, as appearth by his much cost in reparyinge of his Westwood Lodge dwellinge howse, and also in his huntinge that [same] weeke he dyed, and that Satterday was up and though to have rod to his court to Blyborowe in his coarch, but that his friends, wife, and others, perswaded him to the contrary. Hee dyed that night.

Ower poore towne of Walberswick is nowe one of the poorest townes in England, not beinge able to repayer our church, or meatinge place, which at the first was reared up by the inhabitaunts, at their only cost and chardge, and how may poore widowes and fatherless and motherless children, and at this present, not above one man living in the towne that have five pounds per yeare of his owne therein, liveinge.'

The above relation was penned in 1652. (Extracted from the Church Wardens' Account Book in 1754)

Document 14: John Barwick's letter of 15 May 1652. [HA30: 50/22/27.3]

Reverend Sir,

Seeing it hath pleased God to have mouved an honest poor man of our town one James Reynolds of Walberswick, this poore man knowing our oppressions to be great and soe grevious that we live in and are like to conynue. He and another man cam to our house and put us in mynd of drawing out of a petition, and he will take it in hand to present it without any great charge. I told him that we are soe poore; what little money our poore neighbours have towards it: yet myself being moved earnestly by some, my selfe with another man made a beginning and gave him his desire, and drew out a petition, and nothing but the truth what ever our adversary may say to the contrary.

For indeed to bring his own ends about he cares not what himselfe and others say and swear to it. I myself have much experience of him this way. as the last wynter was 12 month he brought a tryall then against me because

I would not give him his unjust demands: his councellor and witnesse did much fouly declare and largly swore against me. Yet through the goodness of God in giveing wisdome to the goodman judge Sir John, who found out the foul wittnesse bearer, that then soe fouly and largly all together strove against me, which if otherwise if he had not byn found out, I had byn then more than half undon. I define I may never forget it, that great wonderfull deliverance of myne.

And soe for our parts we dare not neither can we soe falsefie, he know that we are a poore people and cannot appear so personally our ableness cannot afford it, and he loves to take all advantages, for myne own part I am aged between 3 and 4 score yeares [60–80] and one unfitt to travell. Whereas he saith the minister hath the tythes, that is not soe, for he hath let them to one of his cheif fearmers within our town, who have had them formerly, and very stricktly he then gathered them, and now will do no lesse, but have what he pleaseth. When he was to pay to our minister £20 per annum much adoe had to get it, and at last Parliament charges abaled out of it, soe that through much adoe 2 or 3 years the minister have had the tythes, but tooke them away and letted them to his man that now have them, which through want of meanes that faythfull minister could not but leave us, and now at present we have not any minister resident with us, but are glad much to travell fore it; those that can, some cannot by reason of age which is the more to theire soules greife.

He saith alsoe that he can justly answere that we have as much freedome of commonage as we have had within the memory of man, which is another foul one fore since few old wittnesses are yet liveing, as one poore old blynd man Robert Durrant of Bothy that was in Sir Owen Hoptons dayes our town neat herds boy, in those dayes before the Brookes came to it, and after in Alderman Brookes time that purchased the mannor, those few years he lived after, were not prevented of commoning till after that Sir Robert Brooke came to it, then the poore inhabitants were prevented of feeding their great beasts there in that Pauls Fen that is right against his Westwood Lodge. 12 yeares since we had the last tryall that passed on the town side, when then we had 12 able wittnesses that Judge Littleton would then examine by 6. then we had possession neare 2 years in this our Pauls Fen, till afterwards Sir Robert Brooke tooke it away by violence, and soe now his sonne hold it from us still, and yet he pretend much clearverness from doeing us any injury.

His father was the meanes of 4 men loosing their lives by appointing such a stout fellow to abuse the poore inhabitants cattle as he did, and when there was noe more cattell to abuse in their own common low grounds this stout man cam into our town and fell a quarrelling and fyghting that afterwards he dyed. And Sir Robert soe followed it that 3 after were hanged, and would have hanged more.

Sir Robert then alsoe plowed up great store of the upland common heath where on he had great cropps of rye, and the poore, poore gleaning, and one poore woman got a small handfull of corn out of the gavell* not worth 2*d.*

for which cause Sir Robert caused her poore husband to spend many pounds, and this our young Lord John Brooke goeth on in imitateing his father this way to the full for the last yeare our poore neighbours gleaning in barly harvest this gent was in the field when the poore folkes came to glean barly, and saw them himself most part of the day there, and when neere night came the poore people thought to have carryed home their barly gleanes which they had looked for most part of the day, then came the squire himself and took them agayn, though the poore people pleaded much, seeing they had soe spent the day, they told him if he had but held up his hand against them at the first they would have been gone, yet for all this and much more pleading they could not prevayle, soe cam light home saying they had this day gleaned for the Esquire.

And also a poore man appointed by the esquires farmer to mow and to bring in rushes [reeds] for the esquires house at Walberswick, the rushes were cutt and carted most of them to this house whereon they were layed, the esquire heareing where the man cut them, which most of them upon the common waste salts where the esquire hath a waste salt rushy marsh whereon some were cut, the esquire appeared displeased, the poore man then left fetching of them soe that the remaynder they carryed away, the esquire soe threatening this poore old man to have him into jayle that he went to him telling him that they were laid upon his own house at Walberswick, and he was not paid for the mowing and carting of them, yet that would not serve the turn though he fell down on his knees humbly beseeching him yet no mercy could fynd, loth the old man was in his age to lye in the jayle, he got another man to go to the esquire with the old man's wife and carryed with her 6 shillings, soe through much adoe they at last prevayled for that 6s. and 14 more there was paid to him before that midsommer according to agreement soe he had 20s., would have had 40s.

The last year another man [probably Barwick himself] lost a very good cow bigg with calf through his fervant bad useage of the cattle they fynd on the common heath in dryveing them to Blythburgh pound, and then the owners he made pay what he list before had them out 10s. for a yearling calf, the poore man offered to sell it for 20s., and others likewise extreamly paid before could get them out. This paper cannot conteyn all be the half of his courses however he, now I heare this day setteth forth towards London shall carry himself there for his own ends.

This midsommer he brings a tryall against me for goeing our bounds as we have gone allwayes before, and one we bring against him for our commons. Thus in the behalf of our town most of the seamen at sea, and 2 or 3 of our best well wishers weak and sick in bed that have been desireous to write unto your self with thankfullnes for your paynes in writeing to our adversary for us though he have been since much oppressing us with his whole fflock of sheep now just to our doores where he never had them went before upon our common waste saltes where little comfort for great beastes after his flock of sheep and this he will doe, the shepherd is greaved at it, but he sauth that he must doe as his master commandes.

From Walberswick the 15 of May 1652   John Barwick

[PS.]

I pray good Sir pardon my boldnes at this time in the behalf of our poore town.

Hugh was lately at our poore house in Walberswick and desired to be re-membred.

* ON *giofull* 'liberal' – the free – the loose ears of barley that had not been gathered into sheeves.

Document 15: Walberswick Hearth Tax Returns for 1674. Adapted from: *Suffolk in 1674: being the Hearth Tax Returns*, **Suffolk Green Book No. XI, Vol. 13 (1905), pp. 291–2**

WALBERSWICK.
(Rearranged in alphabetical order.)

Family names which appear in the main text are underlined.
 a = First list of 47 hearths.
 b = Second list of 45 hearths from houses empty 2.5 years.
 c = Third list of 45 hearths 'certified for' as too poor to pay.

Jo. Alderman	2c
James Aldridge	3b (Empty 2.5 years)
Mr Bacon	2b (Empty 2,5 years)
Edward Barfoot	4a
Widow Barfoote	2c
Widow Barfoote	2c
Henry Barnes	3a
Henry Barwicke	2b (Empty 2.5 years)
Widow Bird	1c
Jo. Blowers	3a
Jo. Blowers	2c
Sam. Bond	2a
sam. Bond	4b (Empty 2.5 years)
Wr. Borrowe	2a
Isa Borwood	4b (Empty 2.5 years)
Widow Brewster	2c
Widow Britting	3b (Empty 2.5 years)
Thomas Burnett	2a
Thomas Chapman	1c

Miss Chapman	4a
William Chapman	3a
Mr Roger Cooke	3b (Empty 2.5 years)
Mr Cooper	2b (Empty 2.5 years)
Mich. Crispe	1c
William Crowe	1c
Crowe	1c (Empty 2.5 years)
Thomas Eaarne	2a
James Eade	1c
Edmund Fisher	1c
Widow Fuller	2c
Widow Goodwin	1c
Widow Hackett	2c
Thomas Hall	1c
William Harman	1a
William Harman	3b (Empty 2.5 years)
Thomas Harman	2a
Jo. Hearne	3a
Jo. Hearne	2c
Bymell Herrye	1c
Widow Jones	1c
Isa Kettle	4b (Empty 2.5 years)
Jo. Inties, Bishop	2c
Robert Larwood	2c
John Mason	1c
Henry Mayshead	1c
Thomas Michell	1c
Jo. Miles	1c
Edward Mills	2a
Henry Hills	2a
John Mills	3b (Empty 2.5 years)
Widow Mills	1b (Empty 2.5 years)
Honor Moneys	3b (Empty 2.5 years)
Bar. Mosse	2a
William Neave	2c
Widow Pewes	1c
Widow Preston	4a
Jo. Ravens	1c
Widow Sallows	2c
Widow Smith	2c
Dan. Studly	2a
Jo. Taylor	1c
William Taylor	4a
Robert Thacker	2c
Mr Warren	4b (Empty 2.5 years)

Document 16a: Beating of the Bounds of Walberswick. Thomas Gardner's transcription from the churchwardens' account book (FC185/E1/2)

The Bounds of Walberswick, May 6, 1678.

From the [river]channel cross the marshes, by the further end of Palls Fenn, westward unto the commons, where there are three crosses by the side of the marshes belonging to the house, where now Goodman Hows the shepherd dwells; and from thence, back by the old bank, 'till we come right against the Park-House [Westwood Lodge], going in at the great garden gates, and then right forth cross the park, where there are several trees marked with letters, and marks of our inhabitants, and thence to the maple by the park-side, where there is a cross upon the common, and thence right cross to Deadmans- Cross, where the boys heaved stones to the old heap, according to their old custom. On the meer at Deadman's-Cross (called so by a Man being buried there for suicide) formerly grew an ash-tree, whereon were made hundred of marks by the townsmen of Walberswick, in old times when they went their bounds. This tree was digged up by one Shippens, a poor man, and old Mason, of Blythburgh, by the order of the farmer that then lived at Westwood-Lodge.

1644, May 30 being Thursday, when the townsmen went their perambulation of their bounds, Henry Richarson, one of the antientest and chiefest men in Walberswick, reported at Deadman's-Cross, in the presence of more than forty souls, that he was one of the church wardens when that tree was fetched, by his partner Henry Feardns, from the place where it was digged up, and laid into the Church of Walberswick, where it continued more than twice seven years. And upon a time, when the plumber was at work on the leads, John Arnold being then church-warden (and fireing scarce) gave order to William Thorrold, the sexton, to rive it out for the plumbers use; which the said Thorrold also acknowledged and confessed.

3. May 18, 1683. We went our antient town-bounds, there being about forty in our company, from the west of Pauls-Fenn, and so up the green path before the Lodge, and so through the park to Deadman's-Cross, by the heap of stones. Penned by me, James Dier, Minister.

Document 16b: Beating of the Bounds of Blythburgh. From a document in the Cockfield Hall papers. HA30:50/22/3.1 [55]

Titles per defens to Palls Fen

Sy Austen 50 (*et al.*)

Sayeth that he hath gon perambulation with the cheifest of the towne of Bly [burgh] above 50 years since and hath (usually) taken in Palls Fen within the bounds of Bly [burgh]. And he never knewe it to be partch in by the inhabitants of Walb [erswick] within ther perambulation. And that during the tyme of his reach [memory] which is for 50 years and upwards he hath known the forsaid fen to be heald and occupaed by Sir Arthur Hopton in tymes past Lord of the Mannor of Bly: and since by Sir Robert Brooke his father maker and aymsethl [sic] and ther farmors and that he never knowe any Walb [erswick]s cattel feed there.

They further saith that about 50 years since it was usually overflowen by the sea tydes soe that cattle could not feed there upon untill Sir Robert Brooke['s] father att his great cost and charges made a bank to keepe out the salt water [1597–1601] And the forsaid pales of Westwood [Park] were sett by the syde of Palls Fen soe fare as ther would (by reason) for water.

Tho. Heale: saith that he hath gon perambulation for these 30 years and hath usually taken in Palls Fen with the perambulation of Blyburgh and that Sir Robert Brooke his farmers hath quietly enjoyed the same fore 30 years upwards.

Viz. Lease 16 Eliz [1573/74]. That Palls Fen lyeth in Blythburgh. And the forsaid [leasee?] deceased alledged it to lye in Walber [swick]. Palls Fen contains 26 acres or there aboute.

Index